U0053504

營養進補 坐月食譜

集結中醫師、營養師的補身、瘦身食譜

跨版生活

編　　著：徐思濠（註冊中醫師）
營養顧問：胡美怡（註冊營養師）

目錄

前言 產後美麗大變身

中國人一向很重視產後「坐月」，認為坐月期間調養得好，不但能令女性補充分娩時所失去的，更是改善懷孕前身體出現各種毛病的大好機會。很多長輩常說，坐月調養得宜，是女人一次「脫胎換骨」的好機會。

所謂「坐月」是指分娩後的四至八個星期，醫學上稱之為「產褥期」。因為女性在生產時，身體消耗了大量體力和營養，需要一段時間才能完全恢復。徐思濠醫師認為，在這段期間，最好多臥床休息，配合適當飲食，調養好身體，令身體和生殖器均能盡快恢復健康，對母嬰身心健康均有裨益。有關資料可以參考本書P.14-19的「坐月飲食原則及宜忌」。

然而，當小寶寶呱呱落地後，很多媽媽，尤其是年輕的新一代媽媽，她們關心的不是補身，而是產後瘦身的問題。

原來懷孕期間，由於賀爾蒙的變化，加上進補所吸收的營養，身形會變得肥胖臃腫，因而很多媽媽都會擔心不能恢復懷孕前的苗條身形，有的甚至急於節食瘦身。

其實產後瘦身應由懷孕時就開始，假如在懷孕期間能好好控制體重，產後瘦身就能事半功倍（有關資料可參考

P.26-27「坐月瘦身原則及方法」）。即使懷孕期間體重真的超於標準，當然要減肥，但也不要操之過急，否則自身營養不夠，對坐月期間的身體調理絕對有影響。

至於產後瘦身，原來餵哺母乳是其中一個有效的控制體重方法，因為媽媽每次餵哺母乳大概可以消耗約500大卡路里，比起每天步行20分鐘才能消耗約100大卡路里確是有效得多。

另一個控制產後體重的方法，就是控制卡路里的攝取。據註冊營養師胡美怡小姐表示，我們的體重每下降一磅需要約3,500大卡路里的負差，每天攝取少500大卡路里，一星期就能減一磅，而健康減重的速度是：每星期約減輕1至2磅。但要注意，每天能量攝取不能少於1,000大卡路里，因為除了營養攝取不足外，新陳代謝率亦會減低，減慢復元速度及導致抵抗力下降。這部份可以參考本書P.26-32的「坐月瘦身原則及方法」。

其實無論餵哺母乳與否，坐月期間都應以「健康」作為飲食的大前提，絕對不應該急於實行「節食」，因為產後進補不足除了令母乳營養少外，母親自身的儲備亦會因被提取而使用，增加將來患上如骨質疏鬆症等疾病的機會。

假如你對坐月完全沒有概念，家中又沒有長輩提點，可參考本書兩個分別由中醫師和營養師設計的「30日餐單」，給自己進補一番，算是獎勵自己懷胎十月誕下寶寶之功勞。

第一部份：

坐月飲食
原則及餐單

1. 為甚麼要坐月子？

　　產婦由分娩出胎兒、胎盤後到生殖器官復原需要一段時間來恢復，一般需要42至56天左右（約6-8週），我們稱這段時間為「坐月子」，即醫學上所稱之「產褥期」。在這段期間，產婦最好多臥床休息，配合適當飲食，調養好身體，令身體和生殖器均能盡快恢復，有助母嬰身心健康。

　　古語有云：「女人產後補」，顧名思義「補」亦有分時候。對女性來說，產後進補既是理所當然，又是身體最需要補充的時間，常被人喻為人生的第二度發育，是重塑身體的大好時機。打個簡單的比喻，口乾時喝水功效較不口乾時喝一杯水感覺更滋潤、更解渴，故此要把握機會，不要錯失良機！

坐月等於30天？

　　很多人一聽到「坐月」，便聯想到產後須休息一個月，這觀念是不對的。事實上由懷胎開始，為了孕育胎兒的成長，孕婦身體起着很大的變化，例如盤骨鬆大了、脊骨彎了，當分娩後身體需要一段較長的時間才能完全恢復過來。基本生理恢復要5至6週，再加上每個產婦體質不同，所以坐月只是一個統稱，而不是一定只有30天。

飲食針對「多虛」「多瘀」的特點

　　要有效地調理身體，除了充份休息外，還要注意「坐月子」期間的飲食保健及體格鍛鍊等多方面配合，因應產婦體質「多虛」及「多瘀」的特點，及時調理，免得因失於調理，而留下產後病發生，令日後體質變差。

書內介紹多款食療保健湯方、食方及解釋一些產後常遇問題，希望能協助每位準媽媽食得健康、補得其法。

產後的體質

產婦因分娩時造成產創及出血，令體內元氣受損，抵抗力減弱，故產後有「血脈空虛」的說法，因此產婦在生活各方面均須多加注意。古人則有云：「飲食宜淡泊，勿食生冷堅硬，勿過食肥膩厚味，以免食傷；居室宜避風，衣著需溫涼適宜，以防外感，不宜交合（房事），不宜過於勞役；悲恐憂鬱，皆不可犯」的種種說法。由此可見，產後若因起居飲食不慎則容易引起產後各種疾病。

產後常見的疾病有：

產後發熱、產後出血、惡露不下、惡露不止及乳汁不下等。

要留意產婦的體內情況，主要表現於以下三方面：

1. 小腹有否作痛，以辨惡露尚有沒有；
2. 大便通與不通，以辨體內津液之盛衰；
3. 乳汁行與不行，以作了解消化吸收機能之強弱，也和飲食多少相關。

透過以上三大重點便能對產婦體質了解更深，再進行「虛者宜補，實者宜攻，寒者宜溫，熱者宜清」。還須在食補時考慮到產後情況，以產婦氣血俱虛為本，故宜照顧氣血，不宜過用溫燥，熱亦不宜過用寒涼，以「產後宜溫」為保健食療指導之原則。

2. 解讀10大坐月疑惑

疑惑1 坐月期間真的不能洗頭沖涼，甚至不能沾水嗎？

在中醫角度中並未有「坐月期間不能洗頭和洗澡」的說法，但是有些產婦在坐月期間洗頭、洗澡後又好像對身體有些影響。其實這些禁忌又未必沒有道理，它都是由前人的經驗總結出來，只是未有加以分析，故被人認為不科學。

中醫學理論則認為女性本質屬陰性，大自然中，凡一切寒冷的、下降的、靜止的、機能減退的表現均屬於陰。由於自然界有陰陽對立統一的關係，故生活中亦需作出陰陽平衡之調理方法。

每位產婦於產後氣血虧虛、元氣受損、陽氣不足（臟腑退化機能減弱）、身體正氣下降（抵抗力下降），故體質較為虛弱。因而沾冷水、沖涼、洗頭時容易接觸風寒、水濕，易被外邪侵襲，令產後受感，故不應或減少進行沾水洗澡等活動。

但時至今日，由於現代科學昌明，發明了暖氣機，又有熱水爐、暖水袋、電風筒等個人護理用品，大大減低沾水洗澡時或洗頭時受感的機會。**產後婦女只要縮短在浴室洗頭、洗澡的時間，避免當風，並選擇於日間進行，而且梳洗後盡快用乾布抹身及以風筒吹乾頭髮，便可減低着涼的機會。**但對一些體質偏差、神疲易倦的產婦來說，還是以乾洗或以薑水抹身清潔為宜，避免着涼，因為產婦體虛，若被外邪所傷，引發感冒，治療時間偏長，又影響體質康復，還會影響照顧初生嬰兒，因此應當小心處理。

老人家說，用老薑煲水洗澡，日後不會頭痛，對嗎?

以老薑煲水洗澡有助祛風散寒。由於產婦體質偏虛，皮膚毛孔疏鬆，較易感受外邪。每當感受外邪又易引發頭暈頭痛，故此以薑水洗頭、洗澡有助減少頭痛的機會。

甚麼時候開始進補？有人說坐月分為休養期、小補期及大補期，對嗎？

產婦在產後進補有助身體加快復元，若進補過急，不但未能奏效，還會影響進補效果，所以不應急於進補。按身體復元情況，可以分開以下三個階段進補。

產後1星期可當作為休養期，因生育時氣血流失、正氣減弱，身體未能恢復，過早進補反而會阻礙脾胃吸收，應以清淡飲食為主。待體質逐漸回復、精神好轉後，可作一些簡單的湯水或茶水調理，如黑豆水、南棗茶等。

惡露乾淨後（約20天左右），可逐漸作一些較簡單的進補，例如運用黨參、北芪、紅棗、杞子、首烏、當歸作食材，避免進補過急，使脾胃功能容易接受。

再過1星期後，若身體表現較佳、未有任何燥熱上火反應，可**開始加強進補內容**，如結合一些功效較強的補品，如鹿茸、鹿尾羓、人參、蟲草等，令機體加速改善，爭取在這個吸收效果最好的階段，大補一番，否則「蘇州過後無艇搭」，即使之後再進補，效果不能相提並論。

除了應在不同時期作不同進補方法外，**還應注意自己體質偏向**，如偏熱人士適宜多作滋陽補血，偏寒人士適宜溫陽補胃。錯誤運用方法不能有效調理補身，所以若不清楚自己體質偏向，請向妳熟悉的中醫師請教一下。

疑惑4 有人說，坐月期間不要飲白開水，怕濕氣滯留在體內，造成日後種種身體不適，對嗎？

飲水對每個人的體質尤為重要，飲白開水對身體並沒有甚麼影響，但要**避免飲冷水便可**。飲用溫水有益機體，避免受寒、損傷脾胃。當然飲用一些特別烹調的水，如黑豆水、紅棗水、南棗水、圓肉水等更有助身體回復正氣。

疑惑5 不能吃生冷食物（如魚生、雪糕、凍飲品）嗎？

坐月期間當然不應進食生冷寒涼食物。寒涼食物進入身體時，容易影響脾胃消化功效，直接影響氣血生成，減慢了吸收，妨礙機體康復。還有不論是否生孕過的婦女，要保養美顏，都應減少進食生冷食物，有助養生防病。

疑惑6 可以吃水果（如香蕉、西瓜）嗎？適合吃甚麼水果？

產後可進食一些性質較平和的水果，例如蘋果、香蕉、橙、提子、車厘子、龍眼、奇異果等，但一些性質偏寒涼的水果應避免於產後一個月內（坐月期間）或體質未回復時食用，如西瓜、木瓜、蜜瓜、火龍果、梨、柿等，否則容易引致脾胃受寒，影響吸收，引致體虛、氣血不足、乳汁生成減少或易患感冒等。

疑惑7 可以飲紅酒嗎？

坐月期間飲用適量紅酒，有活血養血的功效，有助血液循環、強壯身體，但不應在惡露乾淨前飲用，避免引致惡露不淨。另外，需注意飲用份量，每次飲用1小杯，不應酗酒，否則得不償失。

每次可飲用一小杯紅酒。

疑惑8　甚麼時候可以吃豬腳薑？

豬腳薑有補血去瘀、滋陽補胃之功效，但不應產後立即進食，**應當在體質漸覺回復、胃口食慾漸增、產後約10天才可進食**。每次進食亦不宜過多，開始時可單獨飲用醋，令脾胃容易接受，然後每餐少量進食便可。飲食過量會引致肥胖，故此應作適量控制，適可而止。

疑惑9　甚麼時候可以吃薑炒飯、煮雞酒等坐月食品？

產婦於產後1星期便可逐漸回復正常的飲食，而薑炒飯、煮雞酒亦可於產後1星期便開始食用，但要注意身體狀況。有些產婦於產後經常飲用雞湯補身，又會飲用補虛強身的燉湯、進食鹿茸及人參等，容易引致積有熱邪，而薑炒飯與煮雞酒皆是溫性食品，所以**每當身熱多汗、手足心熱、睡時盜汗等都應減少進食**，避免引致熱氣上火，否則要飲藥清熱時，之前所作補的食療都會化為烏有。

薑炒飯可於產後
1星期食用。

疑惑10　坐月期間想減肥，可以嗎？應該怎樣減肥？

坐月期間為產婦體養調理的時期，最好不要過於進補調養身子，進補的食材中，難免會有一些營養成份偏高、致肥的食物，過度進食會引致肥胖，但適當控制進食的次數及份量就可以避免增肥。

其實，產婦不應於產後坐月期間立即進行減重計劃，最好當體質恢復、氣血充盛後才進行健康的減肥方法。若然太早開始，除了令身體未能有效回復補充外，還可能令身體更加變弱，得不償失。最有效的減重方法是注意每日進食的攝取量，及多作運動鍛鍊，少食零食。**建議於產後2個月，因應身體狀況才開始減肥計劃**。

3. 坐月飲食原則及宜忌

基本飲食原則

產婦因生產時失血耗氣，令元氣大傷，再加上部份產婦採用人奶餵養嬰兒，故需要補充大量營養，才可補充所流失的氣血及易於早日恢復健康，更有助嬰兒營養充沛、發育成長。

中醫學認為脾臟為「後天之本」，又有「氣血生化之源」之稱。產婦氣血不足、身子虛弱，故必須注意補益脾胃，促使脾胃得以健運，臟腑生機旺盛，令氣血生成來源充足，正氣精血充沛，則元氣得以恢復，故民間一向有產後宜補之說法。除了注意健脾外，還需配合補益肝腎。腎臟為「先天之本」，主理體內生長、生殖、發育之所須，故生育均須耗用腎精，令腎精虛損，腎精不足，無以養肝，肝失所養，又因肝臟主藏血而導致肝血不足，故肝腎不足之產婦常見面色萎黃、筋骨不暢、白髮長生、爪甲不華之體虛表現。

開始進補時，除注意上述要點外，還不應過早選用一些過度肥甘味濃之食品，原因是產後第一週，因產子時體力消耗過多，令脾胃受損，**故產後調補早期應選用一些較清淡而又易消化之食物**，以免脾胃食滯；同時不應進食生冷，以免損傷脾陽，減弱脾胃消化功能。待身體基本康復後，可漸進食蛋白質及其他營養豐富的補充食品。

坐月飲食宜忌

補氣：人參、北芪、靈芝　　　滋陰：花膠、海參、鮑魚
補血：當歸、首烏、紅棗　　　固本：冬蟲草、燕窩
補腎：鹿茸、豬腰、杞子　　　增乳：海鮮、豬蹄、蛋糕
補脾：淮山、圓肉、黨參　　　回乳：麥芽、花椒

增乳之產婦忌食

1. 麥芽、麥芽糖、麥芽製品，有回乳作用，引致缺乳。
2. 禁食寒涼生冷之食物，如：冰水、魚生。
3. 禁食花椒，有回乳作用，食用導致乳汁減少或斷乳。

回乳之產婦忌食

1. 海鮮、河鮮之催乳食物，如：魚、蝦、蟹、鯽魚等。
2. 甜膩食品，如：奶油、蛋糕、豬蹄等。

理論篇

補身篇

瘦身篇

餐單篇

剖腹產飲食宜忌

剖腹產的產婦須特別注意手術前後的飲食調攝。

手術前忌食人參或西洋參：手術前不應過食或濫服人參及西洋參，避免影響手術操作及手術後休息。因人參藥性強烈，有大補元氣之功效，更有強心、興奮的作用，易令手術時刀口容易滲血，或手術後藥力未減，難以休息。

手術後忌食含氣量多的食物：手術後應對傷口的恢復更要多加注意，否則容易影響產婦日後的身體健康及傷口癒合不良等表現。手術後因減輕引起腸脹氣，更應當禁食6小時，6小時後可服用有助疏導排氣之蘿蔔湯，減少腹脹，又能保持大小便通暢。

待產婦排氣後（放屁），可進食一些半流質食物，如爛粥、粉麵、蛋糕等，補充營養，再因應產婦體質逐漸恢復後，才開始正常飲食。直到手術後一星期左右，才作進一步補充營養。菜式多加心思，配搭得宜，使產婦食得開心，助以消化，更能確保母嬰健康。

剖腹手術後不應進食一些含氣量多的食物，如番薯、黃豆、洋蔥等食品，以防腹部脹氣不適。

四季不同注意事項

　　春夏秋冬四時季節，氣候改變，每個季節都有獨特的地方，如春季多風、夏季炎熱、秋季乾燥、冬季寒冷等，由於產婦分娩後耗氣傷血，體質較為虛弱，正氣未足，故特別容易因調攝不當而發生疾病。疾病發生後可引致產婦身體恢復的時間增長，令身心疲累，又容易誘發產後病的發生，影響照顧嬰兒或哺乳及傳染疾病給嬰兒，而嬰兒體質嬌嫩，形氣未充，抗病能力不足，容易引致多種病變，影響嬰兒生長發育等。

　　故順應四時作出調攝，預防疾納，減低發病的機會，採取相應調攝方法，有利產婦康復，為確保母嬰健康作出重要的一步，以下為四季應注意之事宜：

春季多風

　　應多加衣服，避免着涼，亦應穿着尺碼稍大一點的衣服，有助疏氣。室內應常保持空氣流通，減少因衣着過多、散熱不夠、溫度過高，引致產婦中暑情況發生。

　　春天亦為生發的季節，生為生長，發為發育。飲食方面亦可選取一些營養較為充足的食品，補益氣血，助長生肌。

夏季炎熱

　　天時暑熱，不致寒冷，但因產婦體虛亦不應穿衣過少及經常涼冷氣，避免感受風邪，減低感冒傷風的機會。

　　飲食方面不宜因天熱而進食生冷，影響脾胃功能，注意使用補虛食品及藥物。因補虛食品大多偏溫燥，避免過量服食，引致產婦體內燥熱。時常保持身體清潔，出汗後勤換衣裳，以防汗出不潔，引致皮膚疾病發生。

理論篇

補身篇

瘦身篇

餐單篇

秋季乾燥

> 秋季天氣乾燥、氣溫逐漸轉涼，應多加衣被，避免着涼。氣候轉燥容易引致鼻、氣管及皮膚產生不適，故此間常飲用一些滋潤的湯水或糖水如冰糖燉雪耳、椰汁燉雪蛤等，有助滋陰潤燥、生津養顏。

冬季主寒

> 冬季氣候寒冷，外出活動時應注意保暖，避免着涼。寒邪最易減弱人體內的陽熱之氣，引致寒從內生，發為寒病。應減少進食生冷未經煮熟的食物，如雪糕、魚生，避免進一步傷及陽氣，應多進食一些溫陽祛寒的食品，如雞肉、羊肉、熱湯等，但避免食用大辛大熱之食品，如麻辣火鍋，否則傷及胎兒，後果嚴重。

經典坐月必食大解構

豬腳薑

食用豬腳薑，能補虛強壯、有助排出惡露及增加乳汁分泌，為產婦強身補品之一，又為中國民間慶祝添丁的賀喜食品，每逢每戶添丁總有豬腳薑一煲大宴親朋為母嬰祝賀。豬腳薑要在預產期前最少1個月開始煲，煲時要注意：

> （1）薑同甜醋的比例是1比1，即1斤薑用1斤甜醋。薑皮不要浪費，曬乾後密封儲存，留待坐月子時與水同煲滾作洗澡用。
> （2）煲薑醋時煲和薑都要抹乾，不要滲入水份，以免影響薑醋的貯存。
> （3）要準備大瓦煲和小瓦煲各一個，大瓦煲用來煲全部薑醋，小瓦煲食用前用來煲豬腳和蛋。（有關食譜見第46頁）

大瓦煲：用來煲薑醋

小瓦煲：用來煲豬腳和蛋

甜醋

豬腳：
含豐富膠質，能補益氣血，強壯筋骨，有健脾增乳之功效。

薑：
有祛風散寒、健胃止嘔之功效。

甜醋：
有去瘀養血、開胃益食之功效。

薑炒食品

　　薑有助疏風散寒、健胃止嘔，對產後因體虛血弱、風多血少、肢體惡寒的產婦助以溫陽散寒，能紓減不適，更有助脾胃消化，幫助吸收。所以坐月期間無論炒飯、炒菜、煲湯都常配薑同用，例如薑炒飯。（有關食譜見P.50）

坐月期間進食時多配薑同用。

煮雞酒

　　雞酒為舊有社會之平價平民補虛食品，由於舊社會生活樸素，物資未及現今富裕，生活指數低，根本負擔不起購買昂貴的補品，雞有補虛強壯之功，酒亦有行氣血作用，配合其他材料有去瘀生血之功效。在舊時，有雞食已經是難能可貴。現時雞酒亦常作產婦產後常用補益食品之一。（有關食譜見P.52）

糯米酒

　　糯米亦有分普通糯米及紫糯米兩種，而兩種糯米均有補血強壯之功效。其實糯米性質平和，能健脾益胃，補充營養，故能生血，所以用酒浸製，一方面能健脾生血，另一方面又可活血，故此對產後產婦身體虛弱、血氣不足、常見血虛頭暈、面色蒼白、手足冰冷，有改善的作用。

　　糯米酒及酒釀可在各大超級市場買到，但自己釀製亦不麻煩，而釀成後的酒渣（酒釀）更可用來煮雞或糖水，也是坐月進補的佳品。

糯米酒釀法如下：

> （1）1斤糯米加水（剛蓋過米面多少少），放入電飯煲內煮熟。
> （2）把糯米飯攤開，待涼後加入2粒酒餅（如果想酒味濃些可加白酒1瓶，想酒味甜些可加些冰糖），拌勻後放入密封瓶內。
> （3）把密封瓶放在陰涼處貯半個月至1個月即成。

　　這些酒帶點黃色，故又稱黃酒。釀成的糯米酒可以做甜品，如酒釀丸子、酒釀煮蛋（有關食譜見第P.54），亦可用來炒雞，如黃酒煮雞，也是民間常用的坐月食品。

黃酒煮雞

1. 進補食材小百科

鹿茸

　　性溫、味甘咸，有補腎益髓、壯陽健骨、治腎陽不足之功，對精髓虧損所致的虛勞精衰、精血兩虛、腰膝酸痛、畏寒乏力、筋弱神疲、滑精陽痿、眩暈耳鳴、遺尿頻尿的男子、小兒發育不良及婦女崩漏帶下等有幫助。《本草綱目》言其：「生精補髓，養血益陽，強筋壯骨，治一切虛損，耳聾，目暗，眩暈，虛痢。」鹿茸能增強機能活力、消除疲勞、改善食慾和睡眠，更能提高婦女子宮的張力和增強其節律性收縮，促進創傷、骨折和潰瘍的癒合，還促進血紅蛋白組織紅細胞增生。

用量：1錢半至3錢。
中醫師提提你：臟腑燥熱、陰虛火旺及外感表證患者不宜服用。

鮑魚

　　性平味咸，有補氣養血及收斂生肌的作用，能滋陰補肝、益腎明目，多用於陰虛勞嗽，或肝虛目昏暗。現代營養研究，鮑魚含蛋白質、脂肪、碘、鈣、磷等，亦常用於氣血不足的產後乳少及瘡瘍破潰經久不癒等症。

用量：鮮品（九孔鮑）每次放5隻煲湯，或青邊
　　　　鮑每次半隻至1隻。
中醫師提提你：鮑魚為滋補之品，消化不良的人士只喝湯而不吃肉。此外，鮑魚殼為中藥之石決明，具滋陰平肝潛陽之功效，故可一同放進湯中烹煮，增強功效。

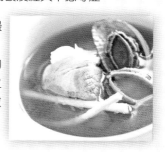

冬蟲夏草

　　性溫味甘，有補肺益胃、止喘咳、補虛損、扶精氣之功，適用於肺胃兩虛、精氣不足、陽痿遺精、咳嗽短氣、自汗盜汗、腰膝酸軟、勞嗽痰血等症。婦人產後氣血不足、腎精虧損、元氣所傷更可常服蟲草，有助固本培元、扶托正氣，迅速恢復體力，補而不燥。

用量： 1錢至3錢（燉湯）或研末服用，每次5分至1錢。

紫河車

　　性溫、味甘鹹，有補腎益精、益氣養血之功。為健康產婦分娩後所出的新鮮胎盤，經剪去臍帶及洗淨附着的血液後，再反覆浸漂而成，古稱胞衣、胎衣等。《本草綱目》言其：「補氣養血、溫腎益精……治虛損疲弱……遺精陽痿，不孕，婦女氣血不足，少乳等。」

《現代實用中藥》言其：「可用於神經衰弱、陽痿、不孕，又促進乳汁分泌。」紫河車，可作藥用或食用，經常被用作燉湯、煮粥、研粉等方法服食。現於中藥店內有售的紫河車為已經炮製過的動物胎盤，使用時更方便和衛生，功效與人的胎盤功效相同。

用量： 1錢。

阿膠

　　性平，味甘，有滋陰潤肺、補血止血之功，適用於陰血不足、婦女月經前後及產後調理。本品補血作用較佳，是各種出血症及治療血虛貧血的要藥，故為婦科的必備良藥。阿膠為動物驢的皮，經漂泡去毛後熬製而成的膠塊，古時以產於東阿（現為山東省東阿縣）而得名，又稱為驢皮膠。

用量： 3錢至5錢，入藥或煲湯，烊化後溫服，或研末後送服。

花膠

　　性平味甘，有補益精血、滋養筋脈、養肝益腎、養血止血之功，適用於腎精不足、崩漏、腰膝酸軟、產後缺乳等症。花膠又名魚肚、魚鰾，為高蛋白滋補佳品，能促進腸胃消化吸收，提高食慾，有利防治食慾不振、厭食、消化不良、腹脹及便秘，又能增加肌肉組織的韌性和彈力，增強體力，又可滋養皮膚，使皮膚更細緻光滑，更能提高機體抵抗力及免疫力，助長發育，增長產婦乳汁生成等功效。

用量：3両至5両
中醫師提提你：花膠含豐富膠質，過量進食容易影響消化功能。

人參

　　性味甘，微苦，有大補元氣、補益脾肺、生津安神之功，常用於氣虛欲脫、短氣神疲、脈微欲絕的危重症候，單獨使用已有效。對脾肺虧虛、心氣不促、氣血虛弱者亦常為滋補要藥。《本草綱目》言其：「治男、婦一切虛症，發熱自汗，眩暈頭痛，反胃吐食，疾瘧，滑世久痢，小便頻數淋瀝，勞倦內傷，中風中暑，痿痺、吐血、下血、咳血、血淋、血崩、胎前產後諸症。」中藥虛方的名稱因其產地及加工製作不同，而有野山參、生曬參、高麗參、吉林參、紅參、白參等稱呼。

用量：1錢半至3錢，切薄片或研粉。每次5分至1錢。
中醫師提提你：人參雖為滋補劑，但若服食不當，亦會產生不良反應。故須留意服用份量及服後感受。遇有服後頭痛、胸悶、煩躁、心跳過速等症狀，請向醫師請教解決方法。

燕窩

　　性平味甘，能滋陰潤肺，益氣補中之功，含有豐富蛋白質、多種氨基酸、維生素、醣類及礦物質、鈣、鉀、硫、磷等。食後容易被消化吸收。近年亦有研究發現，常食燕窩能提高機體免疫力，並具有抗感染、防衰老、防癌抗癌的作用。燕窩宜連續服用，每天空腹服用2至3湯匙已燉熱的燕窩湯更易被人體吸收，比間歇性服用效果更好。

用量：每次3錢作湯料，或用1両先燉好後，分7天食用作保健。

靈芝

　　性溫味甘，有補氣益血、養心安神、止咳平喘之功。《神農本草經》亦有提及靈芝：「益心氣，增智慧，堅筋骨，好顏色，久服，堅身不老延年。」現代藥理研究，靈芝能增強中樞神經系統功能、強心、改善血液循環、增加心肌血氣供應及促進白細胞增加，並有抗過敏、止咳、祛痰作用。

靈芝茯神瘦肉湯

用量：2錢至4錢，切薄片或研粉服。每次5分至1錢。
中醫師提提你：靈芝有不同色澤之分，包括青芝、赤芝、黃芝、白芝、黑芝及紫芝等。常用作入藥或湯料，可選用赤芝或黑芝。

海參

　　性溫，味甘鹹，能補腎益精、養血潤燥，是含高蛋白的滋補食品，《木草從新》提及：「滋陰、補血、健脾、潤燥、調經、養胎、利產。」現代營養研究，海參除了含有豐富的蛋白質，還有合成人體膠原蛋白的原料精氨酸，可促進細胞的再生和機體損傷後的修復，還可提高淋巴細胞的免疫活性，增強人體免疫力，有延年益壽的作用。

用量：1両至3両浸發後作湯料或菜餚食用。
中醫師提提你：海參有滋陰之功，故過量食用則易阻遏脾胃，引致消化不良。

2. 坐月補身10大禁忌

 急於進補

　　剖腹生產的媽媽最好過20朝（惡露乾後）才開始進補。由於很多進補食物都有助補充分娩時所消耗的「血氣」，開刀生產的媽媽為免減慢傷口復原的速度，應慢慢增加膳食裏有助行血壯血食物的數量，例如：紅棗、薑、黨參、薑醋等。

 生冷食物

　　女性產後身體變得虛弱，生冷食物不但有傷脾胃，還有機會令瘀血滯留體內，容易引致腹痛、惡露不止等症狀，所以食物要煮熟後才進食，趁還溫熱的時候進食最好。如果食物放涼了，要再翻熱才好吃。

 刺激性食物

　　產婦不宜進食刺激性的食物如辣椒、大蒜等，這些食物不但影響傷口的復原，而且容易使產婦內熱，引致便秘或痔瘡發作。另外，產婦應盡量避免飲用咖啡及濃茶，以免影響休息，濃茶亦會阻礙鐵質吸收。

 油膩食物

　　女性產後，身體各個器官都需要時間休息復元，進食油膩的食物不但影響身體對其他營養的吸收，還會加重身體的負擔。

 高脂肪、高蛋白質食物

　　坐月進補期間，身體雖然需要補充脂肪和蛋白質，但也要注意吸收的份量，不可一味進食高脂肪、高蛋白質食物，否則容易引致消化不良，更可能使身體吸收過多營養，囤積脂肪，令體重上升。

大麥或麥類製品

　　大麥及其他麥類製品如小麥草、麥芽糖和麥精（奶品）等會減少乳汁的分泌，或出現回奶的情況，所以餵哺母乳的媽媽應避免進食。

 醃製食物

　　醃製食物如榨菜、香腸、臘腸和鹹蛋等大都含有大量鹽份，對一般人的健康而言，多吃也會影響身體健康，媽媽在坐月調理身體期間，就更加要注意不要進食了。

 飲酒

　　酒精能透過母乳傳給嬰兒，嬰兒吸收到酒精後會出現昏睡或遲鈍等情況，嚴重者更會造成神經系統受損，所以餵哺母乳的媽媽要避免飲酒。而傳統的坐月膳食中，有很多含有米酒的食物，因此大家煮食時要注意讓酒精煮至完全揮發才可進食。

 服食營養補充劑

　　現代人經常到餐廳用餐，容易引致營養不均，故很多人都有服食維他命、礦物質或其他營養補充劑的習慣。可是大家需注意「過猶不及」，過量吸收這些營養對身體可能會造成反效果，所以大家只要在日常膳食中做到均衡營養，就可以補充身體流失了的營養。

 半途而廢

　　坐月補品不外乎「紅糖雞」、「麻油雞」、「杜仲花腰」等。最初大家對這些食物可能會有一點新鮮感，但到了坐月期的尾聲，就很容易因為生厭了而放棄。為了自己的健康着想，也得堅持吃下去。

瘦身篇 1. 坐月瘦身原則及方法

（此部份由胡美怡註冊營養師提供）

　　隨着嬰兒呱呱墜地，媽媽們十月懷胎的漫長旅程宣佈結束，但作為媽媽的責任才真正開始。每位產婦在產後的30至40日內都需要休息，讓身體不同部分慢慢復原，俗稱「坐月」。

飲食原則1：懷孕期間應注意卡路里的攝取量

　　「坐月」對中國人傳統來說是大事情，中國人注重食療補身，產後的一個月調理期內就更甚了。可是進補是否等於大吃大喝，或者是高脂肪飲食呢？其實食得多並不代表食得好，進補所指的是飲食均衡，胎兒在腹中的成長與母親的飲食及健康狀況有很大的關連，產後進補多少視乎未懷孕前及懷孕期間的營養攝取是否良好。

懷孕前體重	懷孕期體重增加
過輕（體質指數BMI <18.5）	28 - 40磅
標準（體質指數BMI 18.5-22.9）	25 - 35磅
過重（體質指數BMI >23）	15 - 25磅

*BMI（體質指數）= 體重（公斤）/ 身高^2（米^2）

　　如果大家懷孕期間符合以上的體重增加，產後並不需要特別奉行瘦身計劃。可是很多媽媽仍然很擔心產後身形能否回復苗條。其實我們的身體，體重每下降1磅需要約3,500大卡路里的負差，每天少攝取500大卡路里，一般成年人一星期就能減1磅。

　　在餵哺母乳期間，因為體重超標而需要產後減肥的母親，在整個坐月期內最多只能減2公斤，能量攝取不能少於1,800大卡路里，以助獲得足夠的蛋白質、

維他命及礦物質。攝取過低卡路里除了營養攝取不足外，新陳代謝率亦會減低，減慢復原速度及導致抵抗力下降。

而需要餵哺母乳的媽媽每天所需能量較沒有餵哺的母親多500大卡路里。要攝取500大卡路里其實可以很容易達到，1件芝士蛋糕，或是1盒紙包果汁加1件菠蘿油都能提供500大卡路里。可是，坐月期間為了身體，大家應選擇營養價值較高的食物。

各位媽媽可參考以下卡路里計算表，計算每日從膳食中的卡路里攝取量。

卡路里計算表

食物	簡易份量	卡路里（每100克）	蛋白質／克（每100克）
雞髀（生）（去皮）	1隻	120	19.5
豬扒（生）（去肥）	2塊	140	22
豬腰（生）	½副	100	16
羊腿肉（生）	2片	128	18
龍脷柳（生）	1塊	79	16
蝦肉（生）	15隻	100	20
蛋	2隻	149	12
脫脂牛奶	2/5盒	40	3
木瓜	3/4 碗	40	微量

（資料來源：美國農業部及Nutritionist Five）

飲食原則2：均衡營養

產後飲食選擇合宜能促進身體復原、改善體重並能提供最佳營養給嬰兒。所以每日的膳食應盡量包含不同的養份，包括七大必需元素，即碳水化合物、蛋白質、脂肪、維他命、礦物質、纖維及水。

（1）碳水化合物：提供熱能

　　碳水化合物即是醣質，是提供熱能的主要元素，每天約超過一半的能量都是從碳水化合物而來的。一切含糖或澱粉質食品，如五穀類中的飯、麵、麵包、餅乾；豆類中的眉豆、紅腰豆和花豆；根莖類蔬菜如馬鈴薯、紅蘿蔔；以及水果如橙、香蕉等都含有豐富碳水化合物。

　　身體缺乏碳水化合物會容易出現體力不足、手腳冰冷及情緒不穩等症狀，所以每餐必須有含碳水化合物的食物。當然，大家要注意份量，以免進食超於身體所需而轉化為脂肪。

（2）蛋白質：維持免疫力

　　蛋白質主要功能是組成及修補細胞、荷爾蒙及抗體，有助維持免疫能力。肉類如魚肉、豬肉；蛋類；豆類；硬殼果如花生、核桃；奶類製品如牛奶、芝士都含有豐富蛋白質。

　　蛋白質攝取量不足會令人體重及抵抗力下降，所以無論會否餵哺母乳，媽媽們都不應進食過少蛋白質。可是蛋白質會提供熱能，攝取過多蛋白質同樣會轉化為身體脂肪。

（3）脂肪：產後助你容光煥發

　　很多人都聞「脂」色變，最好所有食物都是低脂或「零脂肪」。其實脂肪對

理論篇

補身篇

瘦身篇

餐單篇

我們的身體十分重要！脂肪主要功能是作為後備能量，保護內臟、滋潤肌膚及頭髮、儲存及運送脂溶性維他命，亦同時維持免疫系統正常及促進嬰兒腦部及視力發展。

餵哺母乳的媽媽們應在坐月期間進食一些含優質脂肪的食物，如沙甸魚、三文魚等，因為有研究顯示它們內裏所含的EPA及DHA有抗抑鬱的功效。

含有優質脂肪的沙甸魚。

身體缺少脂肪會使皮膚乾燥、體重下降及出現缺乏脂溶性維他命的症狀。產後想容光煥發、精神狀態良好，要多進食優質脂肪啊！

（4）維他命

維他命大致分為兩大類：脂溶性維他命及水溶性維他命。脂溶性維他命包括維他命A、D、E及K，能夠溶於脂肪中並由脂肪傳送到身體各部分；而水溶性維他命包括維他命B雜及C，能夠溶於水中，當攝取超於身體所需的時候，會跟隨小便排出。

維他命A：對視覺很重要，可維持免疫功能、有助皮膚及頭髮的生長及維持健康，主要幫助視力發展；媽媽缺乏維他命A會降低抵抗力及增加患上夜盲症的機會。木瓜、南瓜、菠菜、紅蘿蔔等蔬果皆含有豐富的維他命A。

維他命D：維持骨骼和牙齒健康，有助人體吸收鈣及磷。母親如果在懷孕期間鈣質攝取不足，胎兒仍然會從母體吸收足夠的鈣質。所以媽媽們無論產前產後都應攝取足夠的維他命D，以增加鈣質儲備量。人體可從陽光照射下製造維他命D。此外，豬肝、蛋黃、低脂奶及乳類製品都含有豐富的維他命D。

維他命E：屬於抗氧化物，同時幫助組成肌肉、紅血球細胞及其他組織，要恢復皮膚狀態，維他命E又怎能缺少呢？小麥胚芽、全穀或芝麻、綠色蔬菜等都是提供維他命E的重要來源。

維他命K：有助凝血，尤其對剖腹生產的媽媽們，想產後傷口減少出血情況，記得攝取足夠維他命K。綠色蔬菜、肝臟及豆類均含有維他命K。

維他命B雜：其實是多種維他命B群的統稱，大致和身體能量的產生及精神狀態有關，缺乏維他命B雜容易令人疲倦，甚至產生抑鬱，產後絕對要確保維他命B雜攝取足夠。瘦肉、魚、糙米、燕麥、綠葉蔬菜等都含維他命B雜。

維他命C：能增強抵抗力、加速傷口痊癒以及製造骨膠原。產後想快速復原、皮膚恢復彈性，別忘記攝取足夠的維他命C。要注意維他命C特別容易在高溫下受到破壞，所以一般中國老火湯中的維他命C含量其實較低。含豐富維他命C的食物包括橙、西柚、奇異果、士多啤梨、番茄和菠菜。

菠菜含有豐富維他命A和C、葉酸、鈣和鐵，很適合產後的媽媽啊！

（5）礦物質

礦物質是形成頭髮、牙齒、指甲、骨骼、賀爾蒙、紅細胞及酵素的主要物質，以下為常見的礦物質。

鈣質：能強化骨骼及牙齒，有助調節心跳及肌肉收縮的功能不用多講，不想將來身高縮水，產後仍應需多攝取鈣質。

鐵質：是組成紅血球細胞的主要元素，如果懷孕期攝取不足或生產時流血多，產後鐵質的補充能令臉色恢復紅潤。

硒：屬於抗氧化物質，有助防止不飽和脂肪酸氧化及維持免疫系統正常，產後攝取足夠能改善肌肉疲勞現象。

鋅：同樣有助傷口痊癒及維持免疫系統正常，並且能保持味覺及嗅覺靈敏。缺乏鋅會減弱味覺、失去胃口、降低免疫能力，所以應多進食含豐富鋅質的食物。

（6）纖維：減低便秘

纖維是植物食物中人體無法消化的部分，而不是肉中那些咬不爛的物質。媽媽們在懷孕期容易出現便秘情況，引致痔瘡，產後腸道蠕動緩慢會增加痔瘡復發的機會，足夠的纖維能促進腸胃的蠕動及減低便秘。

蔬菜含豐富纖維。

（7）餵哺母乳的媽媽要多喝水！

水分佔身體總重量約五成，有助血液循環、運送營養及帶走廢物。水對餵哺母乳的女士更為重要，因為每天身體製造約24至32安士的母乳，平均每天多飲最少1公升水。水分不足會容易疲勞、便秘及製造母乳不足。

餵哺母乳的媽媽平均每天要多飲約1公升流質飲料。

自然分娩與剖腹生產的媽媽

大部分自然分娩的女士在生產後不需特別營養補充。餵哺母乳的媽媽需注意的是，母乳中的蛋白質及乳糖成分不會因為母親飲食習慣而改變，然而脂肪酸如EPA及DHA的成分、若干微營養素如鋅、硒、碘，以及水溶性維他命特別是維他命B2、C，則會受母親的飲食習慣所影響。餵哺母乳的媽媽除了每天增加攝取500大卡路里外，也應均衡地從食物中攝取各種營養素。

剖腹生產的媽媽也許需要因應手術及麻醉藥對腸道的影響而短暫改變飲食選擇，在首星期額外適量增加蛋白質及能量的攝取，坐月期餘下的三星期則可跟從以上飲食原則。如果既是剖腹生產、又餵哺母乳的女士，首星期每天能量攝取較未懷孕前增加最少500大卡路里。

打算剖腹生產的女士可告知醫生及麻醉師希望產後餵哺母乳，諮詢麻醉藥會否影響母乳餵哺。醫院內的助產士能提供足夠支援予母親，如果不是全身麻醉，產後是可以立時餵哺母乳，不過也許因為傷口或體力關係需要其他人士協助。

美國建議每日營養攝取量

營養素	不用餵哺母乳	餵哺母乳期	營養素	不用餵哺母乳	餵哺母乳期
維他命A	700微克	1,300微克	葉酸	400微克	500微克
維他命D	15微克	15微克	維他命B12	2.4微克	2.8微克
維他命E	15毫克	19毫克	鈣	1,000毫克	1,000毫克
維他命K	90微克	90微克	磷	700毫克	700毫克
維他命C	75毫克	120毫克	鎂	320毫克	320毫克
維他命B1（硫胺）	1.1毫克	1.4毫克	鐵	18毫克	9毫克
維他命B2（核黃素）	1.1毫克	1.6毫克	鋅	8毫克	12毫克
維他命B3（煙酸）	14毫克	17毫克	碘	150微克	290微克
維他命B6	1.3毫克	2.0毫克	硒	55微克	75微克

（資料來源：美國Dietary Reference Intake(DRIs)）

四季坐月要訣及建議餐單

香港四季分明，但食物的選擇未必絕對分明，以前所說的「夏天吃瓜，冬天吃菜」未必完全適用，因為香港食物主要入口而來，現在一年四季都可買到瓜菜，不過仍可注意一些食品選擇事項。

春天

春天正值葉菜如菠菜、白菜、莧菜合宜的季節。農曆新年期間，為了應節，大家可在早餐或下午茶選擇一些賀年食品如蘿蔔糕，代替進食其他營養價值較高的食物，如牛奶麥皮。此外，春天較潮濕，令產後人士情緒更為波動，出現產後憂鬱。日常烹調宜清淡，湯水不宜太濃，可適量放點薑，驅除潮濕及悶的感覺。

此餐單既有傳統賀年食品，亦有坐月必備雞湯，繼續給予媽媽豐富鈣質（高鈣豆漿、乳酪、芝麻、白菜）、鐵質（牛肉、菠菜）、葉酸（菠菜、白菜）及EPA和DHA（三文魚）。鐵質有助產婦補充生產時失去的血液，減少疲倦感，優質的魚油有抗抑鬱功效。此外，蒸蘿蔔糕及薑汁菠菜可用較少煮食油，預留配額給其他餐數。

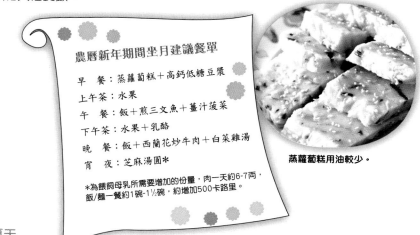

蒸蘿蔔糕用油較少。

農曆新年期間坐月建議餐單

早　　餐：蒸蘿蔔糕＋高鈣低糖豆漿

上午茶：水果

午　　餐：飯＋煎三文魚＋薑汁菠菜

下午茶：水果＋乳酪

晚　　餐：飯＋西蘭花炒牛肉＋白菜雞湯

宵　　夜：芝麻湯圓＊

＊為餵飼母乳所需要增加的份量，肉一天約6-7兩，飯/麵一餐約1碗-1½碗，約增加500卡路里。

夏天

天氣炎熱，媽媽們在坐月期間未必容許逗留在冷氣房間，又被教導不能吃生冷食物，容易食慾不振，如何好呢？

大家日常應多喝流質飲品以補充流汗所失的水分，以免奶水不足。此外，薄荷茶亦有助紓緩悶熱感覺。食品配搭多選擇瓜類，既合季節，水分又多。如果不想吃太多熱的食物，不妨考慮多選擇同樣能提供碳水化合物的生果代替五穀類食物。

有些人會覺得夏天的菜不及冬天的菜般甜美，其實就算不吃綠葉菜，大家同樣能從其他食物中獲得豐富的纖維素。以下建議餐單同樣有必備雞湯，不需用太多煮食油。款式除晚餐外，全部都可以不用熱食，午餐也可以作為沙律般進食。鮮茄所含的鈉質，沒有如罐頭番茄般高，纖維素會較高，而且含有豐富的胡蘿蔔素；雜豆含有鐵質及多種維他命B雜；絲瓜刨鴛鴦皮亦可增加膳食中纖維的含量。

夏天坐月建議餐單

早　餐：麥方包配水果芝士

上午茶：乳酪

午　餐：鮮茄雜豆吞拿魚意粉

下午茶：水果奶昔（低脂奶）＊

晚　餐：飯、絲瓜（駕鴦皮）炒牛柳絲、
　　　　冬瓜（連皮）雞湯

宵　夜：水果

＊為餵飼母乳所需要增加的份量，肉一天約6-7
兩，飯/麵一餐約1碗-1½碗，約增加500卡路里。

水果奶昔。

採用新鮮番茄，
纖維素較高。

秋天

　　秋天早晚溫差較大，是常言進補的季節。這時，大家都很喜歡增加進食果仁、蓮子、百合、銀耳、薏米等食物。此外，秋天剛巧是休漁期的完結，所以亦是大量海鮮供應的好季節。

　　海鮮含有豐富的鋅及優質脂肪，而三米粥既高纖維又低脂，有助平衡進補時攝取過多營養的問題。此外，菜式顏色鮮艷，令秋天加多一點色彩，增加食慾。

秋天坐月建議餐單

早　餐：三米粥（粟米、薏米、麥米）

上午茶：低脂奶＋水果

午　餐：西班牙海鮮炒糙米飯、時菜、
　　　　南瓜湯

下午茶：芝士碎蛋焗薯＊

晚　餐：飯、腰果肉丁、枝竹馬蹄羅漢
　　　　齋、蓮子百合雞

宵　夜：水果

＊為餵飼母乳所需要增加的份量，肉一天約6-7兩，
飯/麵一餐約1碗-1½碗，約增加500卡路里。

果仁含豐富優質脂肪酸。

冬天

　　冬天天氣寒冷，容易減少進食水果和飲水，增加便秘及脫水的機會。冬季時除了注意防寒，亦需注意維他命C的攝取量，攝取不足會增加患流感的機會，影響身體復原進度。

　　此外，很多人喜歡於冬季進食臘味，可是容易令體重增加。與此同時，冬天是菜心、白菜、紹菜、西洋菜等蔬菜的當造季節，不少人都喜歡圍在一起吃火鍋，應確保食物完全熟透才可進食，以免患病，減慢復原的速度及影響餵哺母乳。

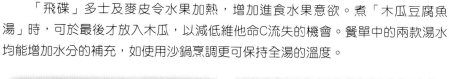

冬天坐月建議餐單

早　　餐：麥方包配水果芝士飛碟
上午茶：熱高鈣豆漿
午　　餐：糙米飯、芥蘭雲耳炒肉片、
　　　　　木瓜豆腐魚湯
下午茶：水果麥皮*
晚　　餐：薑絲菜飯（小棠菜）、
　　　　　蒸銀鱈魚、沙鍋紹菜雞湯
宵　　夜：熱低脂奶

*為餵飼母乳所需要增加的份量，肉一天約6-7両，飯/麵一餐約1碗-1½碗，約增加500卡路里。

　　「飛碟」多士及麥皮令水果加熱，增加進食水果意欲。煮「木瓜豆腐魚湯」時，可於最後才放入木瓜，以減低維他命C流失的機會。餐單中的兩款湯水均能增加水分的補充，如使用沙鍋烹調更可保持全湯的溫度。

糙米飯及菜飯的纖維素含量均較白飯高，
大家應多加選擇。

冬天要注意維他命C的攝取量，
減低患流感的機會。

2. 坐月瘦身10技

① 少吃多餐

女性在生產的過程會消耗大量體力，坐月期間，尤其是生產後頭一星期適宜少食多餐，讓疲累的內臟能夠休息一下。大家可以將一天需要吃的份量分作五至六餐進食，吃完東西最好臥床休息，讓食物慢慢消化。

② 攝取多點纖維素，預防便秘

女性在懷孕期容易出現便秘情況，引致患上痔瘡，產後由於不便於進行劇烈運動，故腸道蠕動會變得緩慢，增加痔瘡復發的機會。每餐攝取足夠的纖維能促進腸胃的蠕動及減低便秘。

③ 注意營養均衡，切勿偏食

很多人都會「聞脂色變」，以為少吃一點脂肪有助減肥，這對於坐月期間的媽媽，尤其需要餵哺母乳的媽媽並不合宜。大家每餐應盡量攝取各樣不同的營養，才能有助恢復體力。為了急於瘦身而不夠營養反而會影響日後的健康。

④ 產後第一週食物宜清淡

太油膩的食物會增加產後腸道的負擔，容易引起消化不良，所以大家除了要注意膳食的營養，還要注意清淡，幫助腸道機能的復原。

⑤ 平衡進補期間的營養攝取量

進補食物的營養價值較高，建議在進補期間，某些餐數可選擇較低脂的食物，以預留配額給進補食物，例如早餐時可進食高纖維、低脂的三米粥（見P.34），以平衡進補時攝取過多的營養。

懷孕時先好好控制體重，為產後瘦身作準備

由懷孕的一刻開始，孕婦便開始展開「進補」的生活。大家在懷孕的時候也應好好控制一下從食物中攝取不同營養的份量，以免吸收過多。如果大家在懷孕時能將體重控制得宜，有助產後回復苗條身形。

餵哺母乳

媽媽餵哺母乳的時候會消耗體內的脂肪，這有助減去坐月期間吃下的脂肪。媽媽只需注意坐月期間攝取足夠的營養，加上餵哺母乳所消耗的脂肪，有助慢慢減去體內脂肪。此外，餵哺母乳還有助促進子宮收縮，有利於媽媽恢復苗條的身材。

不要吃零食

很多人對零食總是難以抗拒，可是零食往往是致胖的主要元兇！女性生產後的頭幾天，甚至整個星期大都被勸喻留在床上多多休息，這些時候，吃零食是最佳的「消遣」活動。可是零食不但往往含有高油量，而且多數含有高鹽，不但容易引致肥胖，而且對身體復原也有所影響。

安排飲食的次序

進食每餐之前，先看看餸菜的內容，決定飲食的次序。蔬菜通常遇到水後會膨脹，容易令人有飽肚的感覺，所以大家可以在吃正餐的時候先喝一點湯，然後吃點蔬菜，再吃飯和餸菜。

少鹽為佳

胃口不佳時容易增加食物中的鹽分以增加口感。不過高鹽分飲食容易引致水腫和增加腎臟的負擔，所以膳食中盡量少鹽及多使用其他天然味料。

1. 中醫30日補身餐單

	早餐	午餐	晚餐
DAY 1	紅棗瘦肉粥	瑤柱豬肝粥	梅菜蒸肉餅、爛飯1碗、蟲草瘦肉燉湯
DAY 2	薑葱魚片粥	馬蹄蒸肉餅、燕窩淮山瘦肉粥(P.72)、蘋果半個	粟米肉片、南棗豬肝補血粥(P.76)、金針木耳排骨湯(P.134)
DAY 3	人參粥(P.70)、烚蛋1隻、益母川芎茶(P.66)、瑤柱豬肝粥	金針雲耳蒸雞、燕窩淮山瘦肉粥(P.72)、清茶1杯	豆腐魚腩煲、白果百合炒絲、薑炒飯1碗(P.50)
DAY 4	火腿蛋三文治、通乳窩蛋奶(P.60)	菜心炒牛肉、薑炒飯1碗(P.50)、香蕉1隻	西芹炒雞柳、果皮蒸泥鯭、薑炒飯1碗(P.50)、黃腳鱲黃豆茨實湯(P.92)
DAY 5	鮮牛肉蛋治、奶1杯	梅菜蒸魚腩、薑炒飯1碗(P.50)、蘋果半個	椰菜煮排骨、杏仁煮南瓜、薑炒飯1碗(P.50)、五指毛桃瑤柱清雞湯(P.98)
DAY 6	牛肉湯意粉、煎蛋1隻、好立克1杯	椰汁炆雞、南棗豬肝補血粥(P.76)、提子5粒	杞子蒸桂花魚、菜心炒肉丸、薑炒飯1碗(P.50)、蟲草燉豬脹(P.88)
DAY 7	牛奶麥皮、烚蛋1隻、益母川芎茶(P.66)	粟米雞翼、南棗豬肝補血粥(P.76)、香蕉1隻	榨菜蒸牛肉、韭王炒鱔糊、薑炒飯1碗(P.50)
DAY 8	人參粥(P.70)、烚蛋1隻、通乳窩蛋奶(P.60)	番茄煮紅衫魚、時菜炒牛肉、燕窩淮山瘦肉粥(P.72)	蒜茸蒸勝瓜、栗子炆排骨、白飯1碗
DAY 9	栗子肉鬆粥、通乳窩蛋奶(P.60)	北菇紅棗蒸雞、白灼清菜、薑炒飯1碗(P.50)、蘋果1個	金銀蛋菠菜、蝦醬蒸豬肉、白飯1碗、杜仲栗子豬脊骨湯(P.100)
DAY 10	菜肉飯、蜂蜜芝麻潤腸茶(P.64)	土魷蒸肉餅、香菇蒸水蛋、薑炒飯1碗(P.50)	魚唇津白煲、枸杞爆肉絲、白飯1碗、百年好合清雞湯(P.124)
DAY 11	菜乾豬骨粥、益母川芎茶(P.66)	牛肉炒豆角、叉燒米粉、橙1個	蒜茸蒸元貝、南乳豬手、白飯1碗、清菜1碟
DAY 12	燕窩淮山瘦肉粥(P.72)、圓肉紅棗茶1杯	銀魚蒸肉片、西蘭花炒蝦仁、香蕉1隻	黃芪灼蝦、豉油雞、油菜1碟、白飯1碗、蟲草黨參瘦肉湯(P.84)
DAY 13	肉片湯烏冬、煎蛋、熱檸茶1杯	煎沙鑽、油菜1碟、白飯1碗、雪梨半個	瑤柱排骨節瓜煲、菜甫蛋角、白飯1碗、天麻鮑魚湯(P.96)
DAY 14	燕窩淮山瘦肉粥(P.72)、炒蛋1隻、通乳窩蛋奶(P.60)	洋葱炒牛肉、白飯1碗、橙半個	薑葱炒魚鱗、腰果西芹雞丁、白飯1碗
DAY 15	芝士火腿三文治、杜仲寄生茶(P.62)	白飯魚煎蛋、清炒糖豆、白飯1碗、清茶1杯	清炒翠玉瓜、枝竹炆火腩、白飯1碗

理論篇
補身篇
瘦身篇
餐單篇

	早餐	午餐	晚餐
DAY 16	雜菌庵列多士、益母川芎茶(P.66)	腐乳蒸雞、蠔油生菜、白飯1碗、蘋果1個	清蒸多寶魚、花生炒豬尾
DAY 17	茄蛋通粉、杜仲寄生茶(P.62)	蒜茸炒白菜、白飯1碗、香蕉1隻	蒜茸開邊蝦、大白菜獅子頭、白飯1碗
DAY 18	雜菜雞絲米、通乳窩蛋奶(P.60)	栗子炒排骨、蝦仁炒蛋、洋葱牛肉、白飯1碗、清茶1杯	荔芋排骨煲、枝竹炆魚腩、白飯1碗、橙1個
DAY 19	牛肉蒸腸粉、艾葉阿膠粥(P.74)	薯仔炆雞、煎鮫魚、蒜茸炒生菜、白飯1碗、橙1個	清蒸黃花魚、節瓜排骨煲、白飯1碗
DAY 20	鮮牛蛋治、通乳窩蛋奶(P.60)	魚肉蝦肉煎蛋角、菜心炒肉片、白飯1碗、香蕉1隻	花膠鴨掌煲、西蘭花鮮魷、白飯1碗
DAY 21	瑤柱雞絲粥、艾葉紅棗茶(P.68)	叉燒炒蛋、豉汁雞翼、白飯1碗、蒜茸西蘭花	茄汁煎蝦、洋葱牛仔骨、白飯1碗、海參淮杞豬骨湯(P.118)
DAY 22	火腿庵列、艾葉阿膠粥(P.74)、通乳窩蛋奶(P.60)	洋葱豬扒、清炒菜心、白飯1碗、提子5粒	洋葱豬扒蒸鱠、魚蓉炒菠菜、白飯1碗
DAY 23	燕窩淮山瘦肉粥(P.72)、烚蛋1隻、杜仲寄生茶(P.62)	薯仔炆排骨、清炒豆角、白飯1碗、香蕉1隻	清蒸石斑、魷魚蒸肉餅、清炒豆角、白飯1碗
DAY 24	栗子肉鬆粥、通乳窩蛋奶(P.60)	排骨炒烏冬、蒜茸炒清菜、橙1個	香菇蒸水蛋、葡汁雜菜煲、洋葱牛仔骨、白飯1碗
DAY 25	瑤柱瘦肉粥、叉燒炒米粉、清茶1杯	梅菜蒸肉餅、魚香茄子、白飯1碗、香蕉1隻	沙薑雞腳、豬腰炒豆角、菜心炒肉片、白飯1碗、充乳湯(P.128)
DAY 26	番茄雞蛋三文治、熱鮮奶1杯	洋葱豬扒、蠔油菜心、白飯1碗、提子5粒	柱侯牛肋條、白切雞、粉絲蝦米雜菜煲、白飯1碗、鹿茸豬蹄湯(P.86)
DAY 27	牛肉炒米粉、炒蛋1隻、艾葉紅棗茶(P.68)	金菇牛肉卷、肉丸湯烏冬、蘋果1個	香草羊扒、炒雜菜、洋葱三文魚頭、茄汁意粉
DAY 28	叉燒湯意粉、烚蛋1隻、熱檸茶1杯	腰果百合炒肉丁、菜心炒牛肉、白飯1碗、香蕉1隻	紅酒燴牛尾、豉油皇煎蝦、白飯1碗、西芹炒魚柳、花膠燉雞湯
DAY 29	南棗豬肝補血粥(P.76)、全蛋麵1個、益母川芎茶(P.66)	冬菇蒸肉餅、魚蛋米線、蘋果1個	粟米斑塊、蘿蔔牛腩煲、蒜茸炒青菜、白飯1碗、參芪魚頭豆腐湯(P.138)
DAY 30	鮮魚片米線、燕窩淮山瘦肉粥(P.72)、艾葉紅棗茶(P.68)	栗子炆雞、時菜炒肉片、白飯1碗	海參翠玉瓜、上湯龍蝦、清蒸鯇魚腩、白飯1碗

2. 營養師10日健美
坐月餐單A（不餵哺母乳）

（此餐單由胡美怡註冊營養師提供）

第1天	早　餐：	麥皮1碗、麥包1片配番茄雞蛋、低脂奶1杯
	上午茶：	果仁1安士
	午　餐：	菠菜雞絲粥（肉2兩）、水果1大個
	下午茶：	高鈣豆漿1杯
	晚　餐：	糙米飯¾碗-1碗、老少平安（鯪魚肉1兩、豆腐半磚）、薑汁時菜
	宵　夜：	水果1大個
第2天	早　餐：	紹菜肉片粥2碗（肉4片）
	上午茶：	水果1大個
	午　餐：	鮮魚片蛋麵（魚片2兩、麵1個）、時菜
	下午茶：	水果1大個
	晚　餐：	菜飯1碗、番茄豬扒、灼時菜
	宵　夜：	──
第3天	早　餐：	高鈣豆漿1杯、麥包芝士吞拿魚飛碟
	上午茶：	水果1大個
	午　餐：	肉粒雜菜湯通粉2碗（肉2兩）
	下午茶：	──
	晚　餐：	糙米飯1碗、蒸銀鱈魚（魚2兩）、炒時菜
	宵　夜：	水果1大個
第4天	早　餐：	營養果仁麥片1碗、低脂奶1杯
	上午茶：	水果1大個
	午　餐：	番茄蘑菇肉醬意粉2碗（肉2兩）
	下午茶：	低脂奶1杯
	晚　餐：	紅豆飯¾碗-1碗、腰果肉丁（肉1兩）、檸檬蒸烏頭配黃芽白（魚2兩）
	宵　夜：	水果1大個
第5天	早　餐：	八寶粥1碗、三絲炒米粉1碗
	上午茶：	高鈣豆漿1杯
	午　餐：	香草三文魚螺絲粉2碗（魚2兩）、時菜
	下午茶：	水果1大個
	晚　餐：	糙米飯¾碗-1碗、西蘭花牛肉（肉2兩）、上湯時菜
	宵　夜：	水果1大個

第6天	早　餐：	麥包芝士吞拿魚飛碟、低脂奶1杯
	上午茶：	—
	午　餐：	紹菜肉片粥2碗（肉2両）
	下午茶：	水果1大個
	晚　餐：	糙米飯¾碗-1碗、三椒炒牛柳（肉2両）、灼時菜
	宵　夜：	水果1大個
第7天	早　餐：	高鈣豆漿1杯、菠菜魚片粥1碗半（魚2両）
	上午茶：	水果1大個
	午　餐：	芝士焗番茄白菌肉醬千層麵1碟（肉2両）
	下午茶：	水果1大個
	晚　餐：	菜飯¾碗-1碗、豆腐羅漢齋（豆腐半磚、豆腐乾半磚）、上湯時菜
	宵　夜：	—
第8天	早　餐：	肉絲雜菜湯米粉1碗（肉1両）
	上午茶：	低脂奶1杯
	午　餐：	生菜雞絲粥2碗（肉2両）、水果1個
	下午茶：	粟米1條
	晚　餐：	紅豆飯¾碗-1碗、蒸銀鱈魚（魚2両）、雙菇扒時菜
	宵　夜：	水果1大個
第9天	早　餐：	營養果仁麥片1碗、低脂奶1杯
	上午茶：	水果1大個
	午　餐：	西班牙炒飯1碗（海鮮3両）、時菜
	下午茶：	—
	晚　餐：	糙米飯¾碗-1碗、麻婆豆腐（豆腐半磚、肉2両）、薑汁時菜
	宵　夜：	水果1大個
第10天	早　餐：	粟米粥1碗半、中芹炒麵2碗
	上午茶：	高鈣豆漿1杯
	午　餐：	香草三文魚天使麵2碗（魚2両）、時菜
	下午茶：	水果1大個
	晚　餐：	糙米飯¾碗-1碗、黃豆炆肉粒（肉2両、黃豆3湯匙）、灼時菜
	宵　夜：	水果1大個

備註：
＊不用餵飼母乳餐單以女士每天需要1,500大卡路里為例，肉一天約5-6両，飯/麵一餐約¾碗-1碗。
＊每天最少飲1.5-2公升流質飲品。
＊低脂湯水隨意。

餐單篇

3. 營養師10日健美 坐月餐單B（餵哺母乳）

（此餐單由胡美怡註冊營養師提供）

第1天	早　餐：	麥皮1碗、麥包1片配番茄雞蛋、低脂奶1杯
	上午茶：	蕃薯1個
	午　餐：	菠菜雞絲粥2碗（肉2-3兩）、水果1大個
	下午茶：	果仁1安士、高鈣豆漿1杯
	晚　餐：	糙米飯1碗-1碗半、老少平安（鯪魚肉2兩、豆腐半磚-1磚）、薑汁時菜
	宵　夜：	水果1大個
第2天	早　餐：	紹菜肉片粥2碗（4片肉片）
	上午茶：	水果1大個
	午　餐：	鮮魚片蛋麵（魚片2-3兩、麵2個）、時菜
	下午茶：	水果1大個
	晚　餐：	菜飯1碗半、番茄豬扒（肉3兩）、灼時菜
	宵　夜：	花生麥米粥
第3天	早　餐：	高鈣豆漿1杯、麥包芝士吞拿魚飛碟
	上午茶：	水果1大個
	午　餐：	肉粒雜菜湯通粉2碗（肉2-3兩）
	下午茶：	茄汁豆焗薯
	晚　餐：	糙米飯1碗-1碗半、蒸銀鱈魚（魚2-3兩）、炒時菜
	宵　夜：	水果1大個
第4天	早　餐：	營養果仁麥片1碗、低脂奶1杯
	上午茶：	水果1大個
	午　餐：	番茄蘑菇肉醬意粉
	下午茶：	芝士飛碟1件、低脂奶1杯
	晚　餐：	紅豆飯1碗-1碗半、腰果肉丁（肉1兩）、檸檬蒸烏頭配黃芽白（魚2兩）
	宵　夜：	水果1大個
第5天	早　餐：	八寶粥1碗、三絲炒米粉1碗
	上午茶：	高鈣豆漿1杯、水果1大個
	午　餐：	香草三文魚螺絲粉2碗（魚2兩）、時菜
	下午茶：	芝士雞蛋麥三文治1份
	晚　餐：	糙米飯1碗-1碗半、西蘭花牛肉（肉3兩）、上湯時菜
	宵　夜：	水果1大個

第6天	早　餐： 上午茶： 午　餐： 下午茶： 晚　餐： 宵　夜：	麥包芝士吞拿魚飛碟、低脂奶1杯 水果1大個 紹菜肉片粥2碗（肉2両）、水果1大個 茄汁豆焗薯 糙米飯1碗-1碗半、三椒炒牛柳（3両）、灼時菜 水果1大個
第7天	早　餐： 上午茶： 午　餐： 下午茶： 晚　餐： 宵　夜：	菠菜魚片粥2碗（魚1両） 水果1大個、高鈣豆漿1杯 芝士焗番茄白菌肉醬千層麵1碟（肉2-3両） 水果1大個 菜飯1碗半、豆腐羅漢齋（豆腐半磚、豆腐乾1磚）、上湯時菜 薑汁蕃薯糖水
第8天	早　餐： 上午茶： 午　餐： 下午茶： 晚　餐： 宵　夜：	肉絲雜菜湯米粉1碗（肉1両） 粟米1條、低脂奶1杯 生菜雞絲粥2碗（肉2両）、水果1個 芝士雞蛋麥三文治1份 紅豆飯1碗-1碗半、蒸銀鱈魚（魚肉2-3両）、雙菇扒時菜 水果1大個
第9天	早　餐： 上午茶： 午　餐： 下午茶： 晚　餐： 宵　夜：	營養果仁麥片1碗、低脂奶1杯 水果1大個 西班牙炒飯（海鮮2-3両）、時菜 芝士飛碟1件 糙米飯1碗-1碗半、麻婆豆腐（豆腐半磚、肉2両）、薑汁時菜 水果1大個
第10天	早　餐： 上午茶： 午　餐： 下午茶： 晚　餐： 宵　夜：	粟米粥1碗、中芹炒麵1碗 果仁1安士、高鈣豆漿1杯 香草三文魚天使麵（魚3両）、時菜 水果1大個、低脂奶1杯 糙米飯1碗-1碗半、黃豆炆肉粒（肉2両、黃豆4湯匙）、灼時菜 水果1大個

備註：
*餐單以餵飼母乳女士每天需要2,000大卡路里為例，肉一天約6-7両，飯/麵一餐約1碗-1碗半。
*每天最少飲1.5-2公升流質飲品，餵飼母乳平均增加1公升。
*低脂湯水隨意。

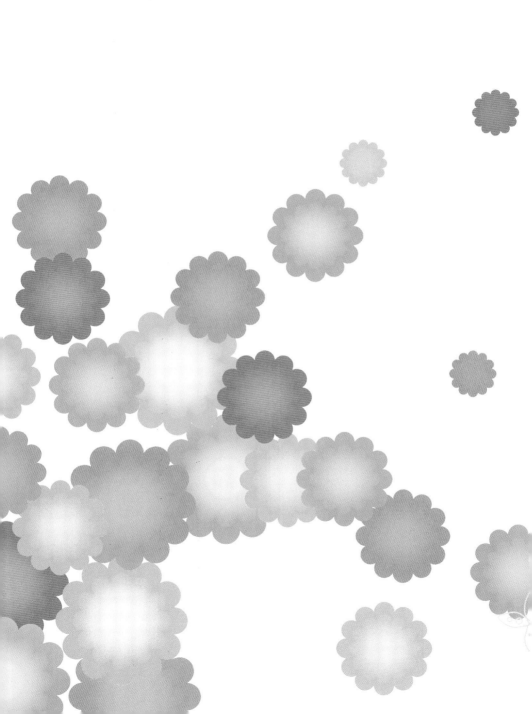

第二部份：

坐月必食及
補身湯水

圍村秘製豬腳薑

Pig's Leg with Ginger

（4人份量 for 4 persons）

適用於：順產、剖腹產、哺乳、
　　　　不哺乳

宜：四季

補：去瘀補血

惡露清：前、後

Ingredients:

6.4kg old ginger, 6.4kg sweet vinegar, 640g black rice vinegar (1 bottle), 1 tbsp coarse salt, suitable amount of pig's legs and boiled eggs

Preparation: (start to cook at least 1 month before the due day)

1. Wash and peel off the skin of the ginger. Dry them under the sun or fry them without using oil.

2. Put the sweet vinegar into a large clay pot. Boil it and add the ginger and coarse salt in. When it boils again, turn the fire off and let it cool down.

3. Boil the same pot of sweet vinegar again the next day. Turn the heat off once when it is boiled. Keep doing the same procedure each day until the baby is born.

Steps:

4. One or two days prior eating, get suitable amount of ginger with sweet vinegar from the large pot to a small pot. Add suitable amount of black rice vinegar. Boil it for a while.

5. Wash the pig's legs and fry them without oil to dry it. Then put them into the small pot.

6. Put in the boiled eggs (without shells) and cook until boils. Turn the heat off and set aside. Eat the pig's legs and eggs the next day.

材料：

老薑10斤、甜醋10斤、黑米醋1斤（1瓶）、粗鹽1湯匙、豬腳適量、熟蛋適量

準備：（預產期前一個月開始煲）

1. 老薑洗淨、去皮，然後曬乾或用白鑊（即沒有油的鑊）炒乾；

2. 將甜醋放進大瓦煲內，用大火煲滾後加入薑、粗鹽，再煲滾後即熄火，放置一旁待涼；

3. 第二天取出再煲滾，一煲滾即熄火，放置一旁待涼，如是者每天取出翻煲至產後。

做法：

4. 食用前一或兩天，從大瓦煲中取適量薑醋，加入適量黑米醋，改用小瓦煲煲滾；

5. 豬腳洗淨，放入白鑊內炒至乾水後，放入小瓦煲內；

6. 放入已去殼的熟雞蛋，煲滾後即可熄火，放置一旁，一天後即可食用。

功效

促進食慾，補充營養，健脾開胃，散瘀止血，強筋壯骨，適合產後氣血不足、手腳冰冷、面色蒼白、神疲怠倦之產婦食用。

Enhance appetite, supplement nutrition, strengthen spleen, remove blood stasis and stop bleeding, strengthen bones and muscles. It is helpful for delivered women who are deficient in Qi and blood, with cold hands and feet, pale faces and tiredness.

中醫師提提您

陰虛火旺及外感表證患者不宜服食，對產婦有補虛損及去瘀之功效，有利惡露排出。

It is not suitable for people who have flaring of fire due to Yin deficiency and syndromes of external contraction. It is also helpful for delivered women to have a tonic for the bodily lost and help to remove blood stasis and discharge lochia.

坐月必食

茶飲

補粥

燉品

補湯

飯餸

坐月必食篇

五更飯
Morning Rice
（1人份量 for 1 person）

適用於：順產、剖腹產、哺乳、
不哺乳

宜：四季

補：驅風健胃

惡露清：前、後

Ingredients:
1 egg, a lot of chopped fresh ginger or shredded ginger, 80g pork

Seasonings:
suitable amount of soy sauce, sugar, corn flour and sesame oil

Steps:
1. Rinse the rice. Cook rice in a rice cooker.
2. Cut the pork into slices. Mix it with suitable amount of soy sauce, sugar, corn flour and sesame oil. Set aside.
3. When the rice starts to boil, add in ginger and pork. Drop an egg onto the surface of rice, cover it and let it cook.
4. Wait 10 minutes after the rice is boiled. It is done.

材料：
雞蛋1隻、大量薑茸或薑絲、豬肉80克

調味料：
適量生抽、糖、粟粉和麻油

做法：
1. 先洗米，放入電飯煲煮飯；
2. 豬肉切片，加適量生抽、糖、粟粉和麻油拌勻，待用；
3. 待飯煮至8成熟（飯面出現火山口狀的泡沫），加入薑和豬肉，打蛋淋上飯面，蓋上煲蓋繼續煮；
4. 飯煮熟後，再焗10分鐘即可食用。

功效

健脾益氣，滋陰強壯，驅風健胃。
Strengthen spleen and benefit Qi, nourish Yin and strengthen the body, anti-rheumatics and strengthen stomach.

中醫師提提您

材料份量可因應個人需要自行調較，首兩個星期可加入木耳絲。
You can adjust the portion of ingredients as you like. You may add slices of auricularia auricular at the first 2 weeks.

薑炒飯
Fried Rice with Ginger
（1人份量 for 1 person）

適用於：順產、剖腹產、哺乳、不哺乳	
宜：四季	
補：驅風健胃	
惡露清：前、後	

Ingredients:
2 bowls rice, 2 egg whites (stirred), 2 tbsp chopped ginger, suitable amount of ginger liquid, 1 stalk Chinese onion (optional)

Steps:
1. Prepare 2 bowls of rice the night before. Put them in the freezer.
2. Chop the Chinese onion finely and set aside.
3. Heat up 1 tbsp oil and Sauté the chopped ginger. Add in a pinch of salt, fry the chopped ginger shortly.
4. Add in the rice and ginger liquid and fry them until the rice separates. When the rice seems like "jumping", put the Chinese onion in and fry for a while.
5. Turn the heat off and add in the stirred eggwhite. It is done.

材料：
飯2碗、蛋白2隻（拂勻）、薑茸2湯匙、薑汁適量、葱1棵（隨意）

做法：
1. 炒飯前一晚準備2碗飯，放在雪櫃冷藏；
2. 葱切粒，備用；
3. 加熱1湯匙油後，爆香薑茸，下鹽略炒；
4. 加入飯和薑汁，炒至飯散開，當每粒飯好像會跳起時，灑葱粒炒勻；
5. 熄火，並加入已拂勻的蛋白即成。

蛋白瑤柱薑粒炒飯
Dried Scallop and Ginger Fried Rice

吃膩了，可以加入瑤柱絲（蒸熟）和芥蘭莖粒（汆水）同炒，成為「蛋白瑤柱薑粒炒飯」，既可轉換口感，亦可增加吸收纖維素。
You may add steamed dried scallop and stems of broccoli (cut into small pieces and blanch for a while). Then it will become "Dried Scallop and Ginger Fried Rice". You can have a new flavor and absorb more cellulose.

功效
祛風健胃，助長食慾，增強體力。
Anti-rheumatics and strengthen stomach, enhances appetite and strengthen the body.

中醫師提提您
外感風熱，內熱煩燥者慎服。
This rice is not recommended for those who have external contraction of wind-cold syndrome, or agitation due to internal heat.

坐月必食
茶飲
補粥
燉品
補湯
飯餸

煮雞酒
Chicken Wine
（1人份量 for 1 person）

適用於：順產、剖腹產、哺乳、不哺乳
宜：四季
補：活血化瘀
惡露清：前、後

Ingredients:
1 tbsp shredded ginger, 1 tbsp shredded auricularia auricular, 1 tbsp shredded Chinese dates, 1/2 chicken, 1 bowl rice wine, 1 bowl glutinous rice wine (refer to P.19)

Seasonings:
suitable amount of brown sugar

Steps:
1. Wash the chicken, peel off the skin and cut into pieces.
2. In a Chinese wok heat up oil and sauté the shredded ginger. Put the chicken in and fry until it dries.
3. Pour the rice wine and glutinous rice wine into the wok. Then add shredded auricularia auricular and shredded Chinese dates. Mix them together, cover tightly and cook for about 10 minutes.
4. Season with some brown sugar and it is done.

材料：
薑絲、木耳絲、紅棗絲各1湯匙、雞半隻、米酒1碗、糯米酒1碗（釀法可參考P.19）

調味料：
紅糖適量

做法：
1. 雞洗淨、去皮、切件；
2. 燒熱油鍋，下薑絲爆香，再下雞塊炒至乾身；
3. 注入米酒和糯米酒，再加入木耳絲、紅棗絲炒勻，蓋上鍋蓋，再煮約10分鐘；
4. 加入紅糖調味即可。

功效
補血養血，活血化瘀，強壯補虛。
Enrich and nourish the blood, promote blood circulation and remove blood stasis, strengthen the body and supplement vacuity.

中醫師提提您
外感患者可待外感康復後才服用。
Those who have external contraction should take it after their bodies are recovered.

酒釀煮蛋

Eggs in Fermented Glutinous Rice

(1人份量 for 1 person)

適用於：順產、剖腹產、哺乳	
宜：四季	
補：行氣活血	
惡露清：前、後	

Ingredients:
2 tbsp fermented glutinous rice, 1 boiled egg, suitable amount of brown sugar

Steps:
1. Put the fermented glutinous rice into the pot. Add water and boil it.
2. Put the egg in and cook for a while. Add some brown sugar and it is done.

材料：
糯米酒釀2湯匙、熟蛋1隻、紅糖適量

做法：
1. 糯米酒釀注入煲內，加水煮滾；
2. 放入熟蛋烹煮片刻，加入適量紅糖即可。

功效

行氣活血，扶正補虛，滋養肝腎。外感患者不宜服用。
Activate Qi and promotes blood circulation, strengthen body resistance and supplement vacuity, nourish and enhance the liver and the kidney. It is not suitable for those who have external contraction.

中醫師提提您

用酒釀來炒雞，加入木耳絲和紅棗絲，也是很好的行氣活血餸菜。
Using the fermented glutinous rice to cook the chicken with shredded auricularia auricular and Chinese dates is a good dish which helps activating Qi and promoting blood circulation.

酒糟雞
Chicken with Distillers' Grain
（2人份量 for 2 persons）

適用於：**順產、剖腹產、哺乳**	
宜：**四季**	
補：**活血去瘀**	
惡露清：**前、後**	

Ingredients:
1/2 chicken, 1 tbsp red distillers' grain, 40g auricularia auricular, 40g Chinese mushrooms, 40g old ginger, 1 tbsp sesame oil, 1 cup water (250ml)

Marinade:
1/2 tsp salt, little red cordial

Seasonings:
1/2 tsp soy sauce, 1 tsp sugar

Steps:
1. Crush old ginger and cut into slices. Soak auricularia auricular and Chinese mushrooms until soft. Drain dry and set aside.
2. Rinse chicken and cut into pieces. Marinate for 20 minutes.
3. Heat pan and add in sesame oil. Stir-fry ginger slices until brown. Sauté distillers' grain until fragrant. Add in chicken pieces and stir-fry until chicken turns golden brown.
4. Add in 1/2 cup water. Cover and stew until sauce is reduced. Add in auricularia auricular, Chinese mushrooms and stir-fry thoroughly. Pour in 1/2 cup water and continue to stew.
5. When the sauce is drying up, drizzle with a touch of red cordial and season with soy sauce and sugar. Cover and cook for a while and it is done.

材料：
雞半隻、紅酒糟1湯匙、黑木耳40克、冬菇40克、老薑40克、麻油1湯匙、水1杯（250ml）

醃料：
鹽1/2茶匙、紅露酒少許

調味料：
生抽1/2茶匙、糖1茶匙

做法：
1. 老薑拍扁後切片；黑木耳及冬菇洗淨浸發好，瀝乾水待用；
2. 雞洗淨切塊後，加入醃料醃20分鐘；
3. 燒熱鍋後下麻油和薑片炒至焦黃，下紅酒糟爆香，加入雞塊拌炒至表面金黃；
4. 加1/2杯水，蓋上鍋蓋轉慢火煮至汁收乾，加入黑木耳及冬菇炒勻，再加入1/2杯水繼續炆煮；
5. 炆至將近收乾汁，潷入少許紅露酒，下生抽及糖調味，再蓋上鍋蓋焗片刻即可。

功效
活血去瘀，扶正補虛，健胃祛風。
Promote blood circulation and remove blood stasis, strengthen body resistance and supplement vacuity, tonify stomach and anti-rheumatics.

紅棗木耳蒸雞腿
Chicken-leg with Chinese Dates and Auricularia Auricular
（2人份量 for 2 persons）

適用於：順產、剖腹產、哺乳、不哺乳
宜：四季
補：活血化瘀
惡露清：前、後

Ingredients:
160g chicken-leg meat, 30g hemerocallis flava, 10g auricularia auricular, 5g fresh ginger, small amount of Chinese onion, 6 Chinese dates (rid of pits)

Seasoning:
1/2 tsp salt, 1/2 tsp sugar, small amount of corn flour, sesame oil and pepper

Steps:
1. Cut chicken-leg meat into pieces. Soak hemerocallis flava, auricularia auricular and Chinese dates until soft and wash. Slice ginger and cut Chinese onion into sections.

2. Add seasoning to chicken meat, hemerocallis flava, auricularia auricular, sliced ginger and Chinese dates. Mix well and marinate. Settle ingredients on a plate.

3. Steam marinated ingredients for 12-15 minutes. Sprinkle with chopped Chinese onion and it is done.

材料：
雞腿肉4両、金針30克、黑木耳10克、生薑5克、葱少許、紅棗6粒（去核）

調味料：
鹽1/2茶匙、白糖1/2茶匙、粟粉、麻油、胡椒粉各少許

做法：
1. 雞腿肉切塊；金針、黑木耳、紅棗浸泡後洗淨；生薑切片；葱切段；

2. 將雞肉、金針、黑木耳、薑片、紅棗加入調味料拌勻醃好，放在碟上；

3. 將醃好的材料隔水蒸熟12至15分鐘，灑上葱花即成。

功效
補血活血，安神強壯，對產後有補虛之功效。
Replenish blood and promote blood circulation, calm the mind and strengthen the body, supplement vacuity at the postpartum period.

中醫師提提您
黑木耳含有豐富鐵質，有助補血。
Auricularia auricular is rich in iron which helps replenishing blood.

坐月必食

茶飲

補粥

燉品

補湯

飯餸

通乳窩蛋奶
Egg Milk for Aiding Milk Flow
(1人份量 for 1 person)

適用於：**順產、剖腹產、哺乳**

宜：**四季**

補：**增乳通乳**

惡露清：**前、後**

Ingredients:

8g rice paper plant pith, 1 cup milk, 1 egg, suitable amount of sugar

Steps:

1. Wash rice paper plant pith and put it into the pot with milk. Boil for 10 minutes.
2. Beat egg with milk, stir thoroughly. Add suitable amount of sugar and it is done.

材料：

通草2錢、鮮奶1杯、雞蛋1隻、砂糖適量

做法：

1. 把通草洗淨，然後與鮮奶一同放進窩中，煲約10分鐘；
2. 雞蛋和奶拌勻後，加適量砂糖調味後飲用。

功效

健脾強壯，通乳增乳，對產後氣血不足、脾虛食少、乳汁不足之產婦有幫助。

Strengthen spleen and the body, increase milk supply and aid milk flow. It is beneficial to the delivered women who are deficient of Qi and blood, spleen vacuity, non-appetite and lack of milk for breastfeeding.

中醫師提提您

餵哺人奶的產婦可每日服一杯，補充身體流失。

Postnatal woman who breastfeed may take 1 cup per day for supplementation.

杜仲寄生茶
Tea with Eucommia Bark and Herba Taxilli
（1人份量 for 1 person）

| 適用於：順產、剖腹產、哺乳、不哺乳 |
| 宜：四季 |
| 補：補肝腎 |
| 惡露清：前、後 |

Ingredients:
12g eucommia bark, 20g herba taxilli, suitable amount of brown sugar, 1 egg

Steps:
1. Rinse eucommia bark and herba taxilli thoroughly.
2. Put eucommia bark, herba taxilli and egg into pot. Add in 5 bowls of water to cook for 45 minutes.
3. Remove the egg shell and add suitable amount of brown sugar. Take after sugar has dissolved.

材料：
杜仲3錢、桑寄生5錢、紅糖適量、雞蛋1隻

做法：
1. 以水洗淨杜仲及桑寄生；
2. 將杜仲、桑寄生及雞蛋放入煲中後，加5碗水煲約45分鐘；
3. 雞蛋煮熟後去殼，加入適量紅糖調味，待糖溶化後即可飲用。

功效
補腎強腰、舒筋活絡、養血柔肝，對產後腰膝酸痛、雙膝無力、面色萎黃、尿頻白髮的產婦有幫助。
Replenish kidney and strengthen waist, relax tendons and activate energy flow in the meridians and collaterals, nourish blood and soften liver. It is beneficial to postnatal women who are suffering from waist and knee pain, knee atony, yellow complexion, frequent urination and greying hair.

中醫師提提您
四季皆宜、藥性平和、可於惡露乾淨前後服用。若血虛氣弱之產婦更可加圓肉5錢一同煎服，增加補血健脾之功效。
This tea is suitable for all seasons and mild in nature. It can be taken before or after lochia. For postnatal women who are suffering from deficiency of blood and Qi, add 20g dried longan to enrich blood and strengthen spleen.

坐月必食
茶飲
補粥
燉品
補湯
飯餸

蜂蜜芝麻潤腸茶
Tea for Moistening Intestine with Honey and Sesame

（1人份量 for 1 person）

適用於：順產、剖腹產、哺乳、不哺乳	
宜：四季	
補：潤腸	
惡露清：前、後	

Ingredients:

suitable amount of honey, 12g sesame

Steps:

1. Stir-fry sesame slightly in wok without using oil until golden brown. Mash into powder.
2. Put sesame powder into a cup. Soak in hot water for 10 minutes.
3. Mix with suitable amount of honey when water becomes warm. Drink it with empty stomach in the morning.

材料：

蜂蜜適量、芝麻3錢

做法：

1. 用白鑊將芝麻略炒至金黃，取出後磨成粉末；
2. 將芝麻粉末置於杯中，加入熱水浸泡10分鐘；
3. 待芝麻水微暖時加入適量蜜糖拌勻，早上起床時空腹飲用。

功效

補氣生血、養陰潤燥、潤腸通便，對產後陰血不足、大腸液虧、血虛便秘產婦有幫助。

Benefit Qi for promoting blood production, nourish Yin and moisten dryness, moisten and relaxing bowels. It is beneficial to postnatal women who are suffering from deficiency of Yin and blood, exhausted large intestine fluids, constipation due to blood deficiency.

中醫師提提您

於惡露乾淨前後均可飲用。每日1至2次，若產婦大便無異常，亦有滋潤之功效，隔天一次便可。

It may be taken before or after lochia and once or twice daily. This can be used as moisturization once every 2 days for postnatal women who have normal bowel movement.

茶飲篇

益母川芎茶
Tea with Motherwort and Hemlock Parsley
(1人份量 for 1 person)

適用於：順產、剖腹產、哺乳、不哺乳	
宜：四季	
補：去瘀	
惡露清：前	

Ingredients:
20g motherwort, 8g hemlock parsley, 1 egg

Seasoning:
suitable amount of brown sugar

Steps:
1. Rinse motherwort, hemlock parsley and egg thoroughly.
2. Put motherwort, hemlock parsley and egg into pot. Add 5 bowls of water and boil until the egg is hard-boiled.
3. Remove egg shell and boil for 10 more minutes. Discard the herbs. Season with sugar and mix together. Eat the egg and drink the tea.

材料：
益母草5錢、川芎2錢、雞蛋1隻

調味料：
紅糖適量

做法：
1. 將益母草、川芎、雞蛋洗淨；
2. 把益母草、川芎、雞蛋放進煲中，加5碗水煲至雞蛋熟；
3. 雞蛋去殼後再煮10分鐘，去渣取汁，略加紅糖調味拌勻，吃蛋喝茶。

功效
益氣養血，活血化瘀，增強體質，對產後惡露不淨之產婦有幫助。
Benefit Qi and nourish blood, promote blood circulation and remove blood stasis, strengthen one's physique. It is beneficial to postnatal women who are experiencing persistent flow of lochia.

中醫師提提您
有化瘀止痛之功，於惡露不淨時飲用最好。
Dissipate blood stasis and relieve pain. It is effective in healing the persistent flow of lochia.

艾葉紅棗茶
Tea with Ai Ye and Chinese Dates
（1人份量 for 1 person）

適用於：順產、剖腹產、哺乳、
　　　　不哺乳

宜：冬季

補：養血溫陰

惡露清：前、後

Ingredients:
12g Ai Ye, 8 Chinese dates (rid of pits), 3 slices fresh ginger, 1 egg

Steps:
1. Put all ingredients into a pot. Add 5 bowls of water and cook until egg is hard-boiled.
2. Remove egg shell and boil for 3 more minutes. Discard the herbs. Drink the tea and eat the egg.

材料：
艾葉3錢、紅棗8粒（去核）、生薑3片、雞蛋1隻

做法：
1. 將藥材與雞蛋一同放進煲中，加水5碗煎煮，直至雞蛋熟；
2. 雞蛋去殼後再煮3分鐘，去渣取汁，飲茶食蛋。

坐月必食

茶飲

補粥

燉品

補湯

飯餸

 功效

溫經散寒、益氣養血、補充營養。產婦可作日常飲料，有強壯補虛、補血祛風之功效。
Warm up channels and disperse cold, benefit Qi and nourish blood, supplement nutrition. Postnatal woman can treat it as daily drink which can strengthen body and supplement vacuity, disperse wind as well as replenish blood supply.

 中醫師提提您

冬季產婦手足不溫、畏寒肢冷者隔日可服一次，惡露乾淨前後皆可服用。
Postnatal women who have cold limbs and feels chilly in winter may take it once every two days. It can be taken before or after lochia.

補粥篇

人參粥
Ginseng Congee
（1人份量 for 1 person）

適用於：順產、剖腹產、哺乳、不哺乳
宜：冬季
補：補益氣血
惡露清：後

Ingredients:
20g ginseng (Korean or Shi Zhu ginseng), 120g rice, 3 Chinese dates (rid of pits)

Steps:
1. Boil ginseng with 8 bowls of water for about 30 minutes.
2. Put in rice and Chinese dates. Cook until mashy to serve.

材料：
人參（高麗參或石柱參）5錢、大米3兩、紅棗（去核）3粒

做法：
1. 先把人參以8碗水煲約30分鐘；
2. 放進大米及紅棗再煲至米爛後即可服用。

功效
大補元氣、健脾和胃、補益氣血、扶正補虛。適合產後氣虛血弱、胃納欠佳、神疲怠倦、面色無華的產婦飲用。
Benefit Qi, strengthen the spleen and regulate stomach, replenish blood and benefit Qi, strengthen immunity function and supplement vacuity. It is suitable for postnatal women who are suffering from deficiency of blood and Qi, poor appetite, fatigue and tiredness, pale complexion.

中醫師提提您
外感表證及內臟燥熱者忌服，冬季天氣寒冷時食用，有助四肢溫照。產後待惡露乾淨後才服用。
It is not suitable for people who have syndromes of external contraction and exterior syndrome, dryness and heat of internal organs. This can help warming limbs during cold winter. It should be taken until lochia stops.

坐月必食

茶飲

補粥

燉品

補湯

飯餸

燕窩淮山瘦肉粥
Congee with Edible Bird's Nest, Chinese Yam and Lean Pork
(1人份量 for 1 person)

適用於：**順產、剖腹產、哺乳、不哺乳**

宜：**四季**

補：**滋陰健脾**

惡露清：**前、後**

Ingredients:
12g edible bird's nest, 16g Chinese yam, 12g Job's tears, 120g rice, 160g lean pork

Seasoning:
a pinch of salt

Steps:
1. Soak edible bird's nest thoroughly. Remove feather and dirt. Set aside.
2. Rinse edible bird's nest thoroughly with other ingredients. Add 10 bowls of water and cook for 1 hour. Season with a pinch of salt and it is done.

材料：
燕窩3錢、淮山4錢、薏米3錢、大米3兩、瘦肉4兩

調味料：
鹽少許

做法：
1. 燕窩浸透後，洗淨去毛及雜質，備用；
2. 燕窩與其他材料一同洗淨後，注入清水10碗，煲約1小時，略加鹽調味即可服用。

功效
有健脾和胃、養血強壯之功效，適合產後體虛、脾胃不和、胃納欠佳之產婦飲用。
Strengthen spleen and warm stomach, nourish blood and strengthen body. It is suitable for postnatal women who are suffering from deficiency of vital energy, disharmony between spleen and stomach as well as poor appetite.

中醫師提提您
放入瘦肉4兩一起煲，使粥更鮮味，健脾滋陰的功效更好。惡露乾淨前後均可服用。
Adding 160g lean pork to cook can make it taste better. The effect on strengthening spleen and nourish Yin will be better too. It can be taken before or after lochia.

艾葉阿膠粥

Congee with Ai Ye and Colla Corii Asini

(2人份量 for 2 persons)

適用於：順產、剖腹產、哺乳、不哺乳	
宜：冬季	
補：溫經養血	
惡露清：後	

Ingredients:

12g Ai Ye, 8 Chinese dates (rid of pits), 8g colla corii asini, 120g rice

Steps:

1. Wash Ai Ye throughly and boil with 10 bowls of water for 1 hour. Discard the herbs.

2. Put rice and Chinese dates in the Ai Ye water and cook until mashy.

3. Add colla corii asini into congee and mix together. Ready to serve.

材料：

艾葉3錢、紅棗8粒（去核）、阿膠2錢、大米3両

做法：

1. 艾葉洗淨後，以10碗水煲約1小時，隔渣取汁；

2. 把大米、紅棗放進艾葉水內煲至米爛；

3. 把阿膠放進粥內拌勻後，即可飲用。

功效

補血養血，溫經止痛，對產後氣血不足、陰虛血少、產後腹痛、面色蒼白的產婦有幫助。

Replenish and nourish the blood, warm channels and relieve pain. It is beneficial to postnatal women who are suffering from deficiency of blood and Qi, Yin vacuity and lack of blood, postpartum abdominal pain, and pale complexion.

中醫師提提您

艾葉的乾品可於中藥店內購買，有溫經散寒之功效。於冬季寒冷時有溫暖手足之功，待惡露乾淨後飲用。

Dry Ai Ye can be bought in Chinese herbs stores. It is good at warming channels, dispersing cold and warming the limbs during winter. It should be taken after lochia stops.

坐月必食

茶飲

補粥

燉品

補湯

飯餸

南棗豬肝補血粥

Congee with Nan Dates and Pork Liver

（3人份量 for 3 persons）

適用於：順產、剖腹產、哺乳、不哺乳
宜：四季
補：養血益肝
惡露清：前、後

Ingredients:

10 nan dates, 80g pork liver, 3 slices fresh ginger, 3 Chinese onions, 120g rice

Seasonings:

suitable amount of oil, salt and soy sauce

Steps:

1. Rinse pork liver thoroughly. Slice and add seasonings. Set aside.

2. Wash nan dates thoroughly. Slice ginger and dice Chinese onions. Set aside.

3. Add rice and 12 bowls of water into pot. Put in nan dates and ginger. Cook for about 45 minutes. Add in pork liver and Chinese onion. Boil until pork liver is completely cooked.

材料：

南棗10粒、豬肝2両、生薑3片、葱3段、大米3両

調味料：

油、鹽、豉油適量

做法：

1. 豬肝沖水、洗淨、切片，加入調味料略調味，備用；

2. 南棗洗淨，生薑切片，葱切粒，備用；

3. 大米連12碗水放入煲中，南棗及薑一同放下，煲約45分鐘後，把豬肝、葱一同放下，煲至豬肝熟透後即可飲用。

功效

健脾和胃，益氣強壯，補血祛風。對產後氣血不足、胃納欠佳、神疲肢倦者有幫助。

Steengthen spleen and warm stomach, benefit Qi and strengthen the body, replenish the blood and disperse wind. It is beneficial to postnatal women who are suffering from deficiency of Qi and blood, poor appetite, spiritlessness and tired limbs.

中醫師提提您

產婦可於產後作簡單食療補充。南棗味道鮮美、調理肝腎、強壯補虛，而豬肝補血滋陰，對產婦均有補益強壯之功。

Postnatal women may take it as an easy supplement. Nan dates are delicious which can regulate liver and kidney, strengthen the body and supplement vacuity. Pork liver can replenish the blood and nourish Yin, which can strengthen the body.

坐月必食

茶飲

補粥

燉品

補湯

飯餸

杞子牛肉粥
Congee with Chinese Wolfberry Fruit and Beef

(2人份量 for 2 persons)

適用於：順產、剖腹產、哺乳、不哺乳	
宜：四季	
補：補中益氣	
惡露清：前、後	

Ingredients:
120g minced beef, 12g Chinese wolfberry fruits, 2 slices ginger(shredded), suitable amount Chinese onions, 80g rice

材料：
牛肉（攪碎）3両、杞子3錢、生薑2片（切絲）、葱適量、大米2両

Seasonings:
Suitable amount of salt and oil

調味料：
鹽、油適量

Steps:
1. Mince beef, season with salt and oil. Shred ginger and dice Chinese onions finely, set aside.
2. Wash Chinese wolfberry fruit and rice thoroughly. Put them into a large pot. Add 8 bowls of water and cook until soft and mashy.
3. Add minced beef, ginger, Chinese onions and mix together. When beef is cooked, season with some salt and it is done.

做法：
1. 牛肉攪碎，以鹽、油調味；生薑切絲，葱切粒，備用；
2. 杞子和大米洗淨後，先放進煲中，加8碗水煲至米爛；
3. 放進牛肉碎、薑、葱拌勻，待牛肉熟透後，略加鹽調味即成。

功效
益氣健脾、補益肝腎、養血祛風。適合產後身倦神疲、元氣虛損、血虛風重、頭重頭暈之產婦食用。
Benefit Qi and strengthen spleen, liver and kidney, nourish blood and expel wind. It is beneficial for postnatal women who are suffering from after-birth fatigue and tiredness, loss of Qi, blood deficiency, and accumulation of wind evils, heavy head and dizziness.

中醫師提提您
四季皆宜，可於惡露乾淨前後飲用，若不吃牛肉之人士可改用豬肉片或雞肉，但以牛肉補中益氣效果較佳。
It is suitable for all seasons and can be consumed before or after lochia. Those who don't consume beef may substitute it with pork or chicken. However, beef are better on tonifying middle and Qi.

燕窩蟲草瘦肉湯
Soup with Edible Bird's Nest, Chinese Cordyceps and Lean Pork
(2人份量 for 2 persons)

適用於：順產、剖腹產、哺乳、不哺乳

宜：四季

補：扶正補虛

惡露清：前、後

Ingredients:
12g edible bird's nest, 8g Chinese cordyceps, 40g lotus seed (without pith and coating), 160g lean pork

材料：
燕窩3錢、冬蟲夏草2錢、蓮子（去蕊、去衣）1両、瘦肉4両

Steps:
1. Soak edible bird's nest in water and rinse thoroughly.
2. Soak Chinese cordyceps in water for 2 hours. Set aside.
3. Put all ingredients into stewing pot. Add 6 bowls of water and stew for about 3 hours. Ready to serve.

做法：
1. 燕窩以清水浸透後洗淨；
2. 冬蟲夏草以清水浸泡約2小時，備用；
3. 把全部材料一同放進燉盅內，加6碗水燉約3小時後即可飲用。

功效
有補益脾胃、扶正補虛、增強體質之功效。對產後氣血不足或病後體虛、及虛不受補的婦女為較理想的滋補食療湯。
Invigorate spleen and stomach, strengthen immunity function, supplement vacuity and strengthen our body. It is beneficial to postnatal women who are suffering from blood and Qi deficiency and weak constitution after prolonged illness. This is an ideal supplement for women who are failure to invigorate due to asthenia.

中醫師提提您
此燉湯藥性平和，不寒不燥。四季皆宜服用，惡露乾淨前後均可服用。
The soup is mild in nature, neither cold nor dry. It is good for all seasons and can be taken before or after lochia.

當歸天麻魚頭湯
Soup with Chinese Angelica Root, Tall Gastrodia Tuber and Fish Head
（3人份量 for 3 persons）

適用於：順產、剖腹產、哺乳、不哺乳	
宜：四季	
補：養血袪風	
惡露清：後	

Ingredients:
12g whole Chinese angelica root, 12g tall gastrodia tuber, 8g hemlock parsley, 5 Chinese dates (rid of pits), 4 slices fresh ginger, 8g angelica dahurian root, 1 big fish head

Steps:
1. Wash the big fish head thoroughly. Set aside.
2. Wash other ingredients.
3. Put all ingredients into stew pot together. Add 8 bowls of hot water. Stew for 3 hours and it is done.

材料：
全歸身3錢、天麻3錢、川芎2錢、紅棗5粒（去核）、生薑4片、白芷2錢、大魚頭1個

做法：
1. 大魚頭洗淨，備用；
2. 其他材料分別洗淨；
3. 全部材料一同放進燉盅內，加8碗熱水，隔水燉3小時後即可飲用。

功效
補血活血、通竅止痛。適合產後血虛風重、頭痛頭暈、面色蒼白、氣血不足的產婦飲用。
Replenish the blood and promote blood circulation, smooth meridian and ease pain. It is beneficial to postnatal women who are suffering from postpartum blood deficiency, accumulation of wind evils, heavy head and dizziness, pale complexion, insufficiency of Qi and blood.

中醫師提提您
此燉湯宜於惡露乾淨後飲用。陰虛內熱或外感患者慎用。
It is suitable to take this soup after lochia has stopped. This should be cautiously used in cases of having internal heat due to Yin deficiency or syndromes of external contraction.

蟲草黨參瘦肉湯
Soup with Chinese Cordyceps, Dang Shen and Lean Pork
(2人份量 for 2 persons)

適用於：順產、剖腹產、哺乳、不哺乳	
宜：四季	
補：補益氣血	
惡露清：前、後	

Ingredients:
8g Chinese cordyseps, 40g dang shen, 3 Chinese dates (rid of pits), 160g lean pork

Steps:
1. Rinse Chinese cordyseps throughly and soak for 2 hours. Set aside.
2. Put all ingredients into stew pot with 6 bowls of water. Stew for about 3 hours. Ready to serve.

材料：
冬蟲夏草2錢、黨參1両、紅棗3粒（去核）、瘦肉4両

做法：
1. 冬蟲夏草洗淨後，浸約2小時，備用；
2. 把材料一同放進燉盅，加水6碗，燉約3小時後即可飲湯吃肉。

功效
滋陰補氣，養血強壯，健脾開胃，有補虛培元、添精生髓之功效。適合產後氣血兩虛、肝腎不足、元氣虛損的產婦服用。
Nourish Yin and replenish Qi, nourish blood and strenghten the body, strengthen spleen and increase appetite, tonify vacuity and nourish genuine essence, replenish essence and generate bone marrow. It is beneficial to postnatal women who are suffering from deficiency of blood and Qi as well as deficiency of liver and kidney.

中醫師提提您
此湯補而不燥，每星期可服2至3次。
This soup can replenish without causing dryness. It can be taken 2 to 3 times per week.

鹿茸豬蹄湯
Soup with Deer Antler and Pork Knuckle

(2人份量 for 2 persons)

適用於：順產、剖腹產、哺乳、不哺乳
宜：冬季
補：強壯增乳
惡露清：後

Ingredients:

12g deer antler, 20g astragalus membranaceus, 4 nan dates, 1 pork knuckle, 2 slices fresh ginger

材料：

鹿茸3錢、北芪5錢、南棗4粒、豬蹄1隻、生薑2片

Seasoning:

a pinch of salt

調味料：

鹽少許

Steps:

1. Wash pork knuckle thoroughly and get rid of the hair. Blanch in boiling water. Set aside.
2. Put all ingredients into stew pot and add 6 bowls of water. Stew for about 3 hours. Season with a pinch of salt and it is done.

做法：

1. 豬蹄洗淨後，去毛汆水，備用；
2. 全部材料一同放進燉盅內加水6碗，燉約3小時後，略加鹽調味即可飲用。

功效

補腎溫陽，補氣益血，扶正強壯，對產後腎陽不振、小便頻多、腰膝酸軟無力、乳汁不足、缺乳之產婦有幫助。
Replenish kidney and warm Yang, replenish Qi and tonify blood, strengthen immunity function. It is beneficial to postnatal women who are suffering from devitalization of kidney and yang, micturition, waist and knee atony and insufficiency of breast milk and lack of breast milk.

中醫師提提您

冬季服用有暖身之功效，尤以四肢不溫之產婦更為適用，待惡露乾淨後飲用。
This soup can keep the body warm when consumed in winter. It is especially good for postnatal woman with cold limbs. It should be taken after lochia is cleared.

坐月必食

茶飲

補粥

燉品

補湯

飯饌

蟲草燉豬脹
Stewed Shin of Pork with Chinese Cordyceps
（3人份量 for 3 persons）

適用於：	順產、剖腹產、哺乳、不哺乳
宜：	四季
補：	健脾扶正
惡露清：	前、後

Ingredients:
8g Chinese cordyceps, 20g lily bulb, 20g Chinese yam, 160g shin of pork

Seasoning:
a pinch of salt

Steps:
1. Soak Chinese cordyceps thoroughly in water for about 2 hours.
2. Put all ingredients into stew pot. Add 8 bowls of hot water. Stew for about 2¹/₂ hours. Add some salt and it is done.

材料：
冬蟲夏草2錢、百合5錢、淮山5錢、豬脹4兩

調味料：
鹽少許

做法：
1. 蟲草先以開水浸透約2小時；
2. 全部材料一同放進燉盅內，加熱水8碗，燉約2¹/₂小時後，略加鹽調味後即可飲用。

功效
滋補強壯，益肺健腎、扶正補虛。適合產後肺虛氣弱、腰酸膝軟、汗多惡寒、面色晃白之產婦飲用。
Nourish and strengthen the system, tonify lungs and strengthen kidney. Strengthen immune function and replenish vacuity. It is beneficial to postnatal women who are suffering from lung vacuity and weak Qi, waist and knee atony, excessive sweat, aversion to cold and pale complexion.

中醫師提提您
體質虛弱、虛不受補之產婦特別適用，四季皆宜，每星期可飲用1至2次。
It is especially good for postnatal women who have weak constitution and failure to invigorate due to asthenia. It is good for all seasons and can be taken once or twice a week.

坐月必食

茶飲

補粥

燉品

補湯

飯餸

太子參薏米豬肚湯
Soup with Heterophylly Falsetarwort Root, Job's Tears and Pig's Stomach
（3人份量 for 3 persons）

適用於：順產、剖腹產、哺乳、不哺乳

宜：四季

補：利水去腫

惡露清：前、後

Ingredients:

20g heterophylly falsetarwort root, 80g raw Job's tears, 80g cooked Job's tears, 5 Chinese dates (rid of pits), 1 pig's stomach, 3 slices fresh ginger, 8g dried tangerine peel, 2 blocks Chinese onions

Steps:

1. Wash pig's stomach with coarse salt and corn flour for several minutes. Rinse thoroughly with water. Blanch in boiling water with ginger and Chinese onions. Let it dry and slice it.

2. Add 12 bowls of water with all ingredients and cook for about 2 hours. Add some salt and it is done. Consume both soup and meat.

材料：

太子參5錢、生熟薏米各2両、紅棗5粒（去核）、豬肚1個、生薑3片、陳皮2錢、葱2段

做法：

1. 豬肚以粗鹽及豆粉洗擦數分鐘，用水沖至乾淨後，再以薑葱汆水，瀝乾、切件；

2. 全部材料加水12碗，煲約2小時，加鹽調味即可飲湯吃肉。

功效

健脾行氣，利尿去濕，適合產後氣血不足、脾虛濕盛、小便不暢、水腫腳腫之產婦飲用。

Strengthen spleen and activate the flow of Qi, promote fluid secretion and dispel dampness. It is beneficial to postnatal women who are suffering from deficiency of blood and Qi, spleen vacuity and excessive dampness, impeded urination and swelling feet.

中醫師提提您

四季皆宜，產婦惡露乾淨前後皆可飲用。每星期可作1至2次飲用。太子參有健脾益氣之功，藥性平和，老少皆宜。豬肚有助利水，薏米有健脾去濕利尿之功效。

This soup is good for all seasons. It can be taken before or after lochia and once or twice per week. Heterophylly falsetarwort root can strengthen spleen and benefit Qi. It is mild in nature and good for all ages. Pig's stomach helps promoting body fluid. Job's tears can strengthen spleen, dispel dampness and promote fluid secretion.

坐月必食

茶飲

補粥

燉品

補湯

飯餸

黃腳鱲黃豆茨實湯
Soup with Yellowfin Sea Bream, Soy Beans and Euryale Seeds
(4人份量 for 4 persons)

適用於：順產、剖腹產、哺乳、不哺乳

宜：四季

補：開胃健脾、增加乳汁

惡露清：前、後

Ingredients:
1 yellowfin sea bream, 2 slices ginger, 80g soy bean, 80g euryale seed, 2 blocks Chinese onions, 3 Chinese dates (rid of pits)

Seasoning:
a pinch of salt

Steps:
1. Remove the scales, gills and internal organs of the yellowfin sea bream. Fry slightly with ginger and small amount of oil. Set aside.
2. Wash soy bean, euralye seed and Chinese dates thoroughly.
3. Add in soy bean, euralye seed and Chinese dates. Cook for about 1 hour with 12 bowls of water. Then add in yellowfin sea bream and cook for another 30 minutes. Add some salt and it is done.

材料：
黃腳鱲1條、生薑2片、黃豆2両、茨實2両、葱2段、紅棗3粒（去核）

調味料：
鹽少許

做法：
1. 黃腳鱲去鱗、鰓及內臟，落鑊以生薑、少油略煎，備用；
2. 黃豆、茨實及紅棗洗淨；
3. 加入黃豆、茨實及紅棗，以12碗水煲約1小時後，放進黃腳鱲，再煲30分鐘後，略加鹽調味即可飲用。

功效

健脾和胃，增強營養，增加乳汁。適合產後乳汁不足、氣血兩虛、肌肉瘦削之產婦飲用。
Strengthen spleen and regulate stomach, enrich nutrition and aid milk flow. It is beneficial to postnatal women who are suffering from insufficiency of breast milk, deficiency of blood and Qi as well as lack of muscle.

中醫師提提您

黃腳鱲為常用保健食療之鮮魚，魚質鮮美、蛋白質豐富、易於吸收。
Yellowfin sea bream is frequently used for maintaining good health. Fish meat is delicious, rich in protein and easy to absorb.

坐月必食

茶飲

補粥

燉品

補湯

飯麵

健脾養血黃鱔湯
Soup for Strengthening Spleen and Nourish Blood with Yellow Eel
（3人份量 for 3 persons）

| 適用於：順產、剖腹產、哺乳、 |
| 不哺乳 |
| 宜：四季 |
| 補：補血強壯 |
| 惡露清：前、後 |

Ingredients:

40g dang shen, 12g Chinese wolfberry fruit, 5 Chinese dates (rid of pits), 3 slices fresh ginger, 1 yellow eel, 2 blocks Chinese onions

Seasoning:

a pinch of salt

Steps:

1. Wash yellow eel thoroughly. Remove bones and internal organs. Blanch with ginger and Chinese onions in boiling water. Cut into pieces. Set aside.

2. Wash other ingredients. Add 10 bowls of water and cook for about 1 hour.

3. Put in pieces of yellow eel and cook for another 45 minutes. Add some salt and it is done.

材料：

黨參1両、杞子3錢、紅棗5枚（去核）、生薑3片、黃鱔1條、葱2段

調味料：

鹽少許

做法：

1. 黃鱔洗淨去骨及內臟，以薑葱氽水去潺，切段，備用；

2. 其他材料洗淨，加10碗水煲約1小時；

3. 放入已切件的黃鱔再煲約45分鐘後，略加鹽調味即可飲用。

功效

補氣補血，滋陰補腎，強壯體質。適合產後氣血兩虛、面色蒼白、血虛風重、頭目眩暈等產婦飲用。

Replenish blood and Qi, nourish Yin and replenish kidney, strengthen immunity function. It is beneficial to women who are suffering from deficiency of blood and Qi, pale complexions, blood vacuity and accumulation of wind evils, dizziness.

中醫師提提您

可以用黃鱔作杞子焗飯，做法簡單（參考P.148），可保健強壯、補氣益血。黃鱔有補血、健脾、強壯之功效，可常作補虛之食療。

Yellow eel and Chinese wolfberry fruit may also be used in rice baking (See P.148). It can replenish blood and Qi, and strengthen immunity function. Yellow eel can replenish blood, strengthen spleen and strengthen the body, which can be used for supplement vacuity in daily dishes.

天麻鮑魚湯

Soup with Tall Gastrodia Tuber and Abalone

(4人份量 for 4 persons)

適用於：順產、剖腹產、哺乳、不哺乳
宜：四季
補：養陰祛風
惡露清：前、後

Ingredients:

12g tall gastrodia tuber, 5 fresh small abalones, 20g Chinese wolfberry fruit, 3 slices fresh ginger, 320g shin of pork

材料：

天麻3錢、鮮鮑魚仔5隻、杞子5錢、生薑3片、豬脹半斤

Steps:

1. Use brush to thoroughly wash abalone until clean. Set aside.
2. Wash other ingredients separately. Set aside.
3. Put all ingredients into a pot. Add 12 bowls of water and cook for about 2 hours. Ready to serve.

做法：

1. 鮮鮑魚用刷清洗乾淨，備用；
2. 其他材料分別洗淨，備用；
3. 全部材料一同放進煲中，加12碗水，煲約2小時即可飲用。

功效

滋陰補腎，養血柔肝，祛風止痛。適合產後肝腎不足、頭暈目眩及腰膝酸軟無力之產婦飲用。

Nourish yin and replenish kidney, nourish blood and soften liver, dispel wind and relieve pain. It is beneficial to postnatal women who are suffering from deficiency of liver and kidney, dizziness as well as waist and knee atony.

中醫師提提您

血壓高者飲用此湯有助降壓。每星期可服1至2次。

This soup helps to reduce high blood pressure. It can be taken once or twice per week.

五指毛桃瑤柱清雞湯

Chicken Soup with Ficus Simplicissima and Dried Scallops

(4人份量 for 4 persons)

適用於：順產、剖腹產、哺乳、不哺乳	
宜：四季	
補：增強抵抗力	
惡露清：前、後	

Ingredients:

40g ficus simplicissima lour, 5 dried scallops, 3 slices fresh ginger, 1 chicken

Steps:

1. Peel off the skin of chicken and get rid of the organs. Wash it and cut into pieces. Set aside.

2. Wash all other ingredients separately. Set aside.

3. Put all ingredients into a pot. Add in 15 bowls of water. Cook for about 3 hours and it is done. Consume both soup and meat.

材料：

五指毛桃1両、瑤柱5粒、生薑3片、雞1隻

做法：

1. 光雞去內臟、洗淨，切件備用；

2. 其他材料分別洗淨，備用；

3. 全部材料一同放進煲中，加15碗水，煲約3小時後即可飲湯吃肉。

功效

補氣強壯，開胃益食，對產後體質虛弱、手足常冷、易患感冒、胃納欠佳之產婦可常作保健飲用。

Replenish Qi and strengthen the body, enhance appetite and promote digestion. It can act as a regular soup for women who are weak after giving birth, with cold hands and feet, easy to catch cold, or with poor appetite.

中醫師提提您

四季皆宜，陰虛火旺、暗瘡口瘡頻生之產婦亦可加西洋參3錢同煲，有清補之功效。

This soup is good for all seasons. People with flaring fire due to deficiency of Yin, having pimple or aphtha frequently may add 12g American ginseng to help cleaning and tonicing the body.

杜仲栗子豬脊骨湯
Soup with Eucommia Bark, Chestnut and Pig's Backbone

（4人份量 for 4 persons）

適用於：順產、剖腹產、哺乳、不哺乳
宜：四季
補：補腎強腰
惡露清：前、後

Ingredients:
20g eucommia bark, 16g cibotium barometz, 15 chestnuts, 20g dried longan, 5 nan dates, 2 pieces pig's backbone

材料：
杜仲5錢、金狗脊4錢、栗子15粒、圓肉5錢、南棗5粒、豬脊骨2段

Steps:
1. Wash the pig's backbones; blanch in hot water for a while. Then let it dry for later use.
2. Wash all other ingredients separately, and get rid of the shell and skin of the chestnuts.
3. Put all ingredients into the pot with 15 bowls of water. Boil for about 2 hours. Add some salt and it is done.

做法：
1. 豬脊骨洗淨，汆水、瀝乾備用；
2. 其他材料洗淨，栗子去殼及衣；
3. 把全部材料一同放進煲中，加15碗水，煲約2小時，略加鹽調味即可飲用。

功效

補腰固腰、強筋壯骨。對產後肝腎不足而引致腰膝酸軟、尿頻夜尿、神疲健忘等產婦有幫助。

Replenish and support for the waist, build up the bones and muscles. It is helpful for delivered women whose liver and kidney are not functioning well, so they feel aching and limp at the back and the knees, frequent urination and night urination, and tiredness and forgetfulness, etc.

中醫師提提您

消化功能差者可加陳皮3錢，栗子煲湯後不須食用。

Those who are weak in digestion can add 12g tangerine peel, but do not eat the chestnut.

花膠雞湯
Chicken Soup with Dried Fish Maw
(4人份量 for 4 persons)

適用於：順產、剖腹產、哺乳、
　　　　不哺乳
宜：四季
補：滋陰補虛
惡露清：前、後

Ingredients:
40g dried fish maw, 1 chicken, 5 Chinese mushrooms, 8g tangerine peel, 3 slices fresh ginger, 2 blocks Chinese onions, 4 dried scallops

Seasoning:
a pinch of salt

Steps:
1. Soak dried fish maw in water until softened. Blanch in water for a while with ginger and Chinese onion. Take out and set aside.
2. Soak Chinese mushrooms and dried scallops in water for 30 minutes.
3. Wash the chicken, peel the skin off and get rid of the fats. Set aside.
4. Put all ingredients into the pot with 15 bowls of water. Boil for about $2^1/_2$ hours. Add some salt and it is done.

材料：
花膠2両、雞1隻、花菇5朵、陳皮2錢、生薑3片、葱2段、瑤柱4粒

調味料：
鹽少許

做法：
1. 花膠浸透，以葱薑汆水備用；
2. 花菇及瑤柱浸泡30分鐘；
3. 光雞洗淨，去皮及脂肪，備用；
4. 全部材料一同放進煲中，加15碗水，煲約2½小時後加鹽調味即可飲用。

功效
滋補強壯、補益精血。適合產後體虛、腰膝酸軟、氣血不足之產婦飲用。
Moisten and strengthen the body, tonify and replenish blood and body essence. It is suitable for delivered women who are bodily infirm, feel aching and limp at the back and the knees, and deficiency of Qi and blood.

中醫師提提您
外感表證患者不宜飲用，脾虛消化功能欠佳者可喝湯而不吃肉，亦可用作燉湯。
This soup is not suitable for those who have syndromes of external contraction. People who have spleen infirmity and weak digestion can drink the soup but do not eat the meat. This soup can also be done in a stewed way.

人參蓮子豬肚湯

Soup with Ginseng, Lotus Seed and Pig's Stomach

（4人份量 for 4 persons）

適用於：順產、剖腹產、哺乳、不哺乳	
宜：四季	
補：健脾去濕	
惡露清：前、後	

Ingredients:

20g ginseng (Korean ginseng or Shi Zhu ginseng), 80g lotus seed (without pith), 1 pig's stomach, 40g hyacinth beans

Steps:

1. Cut the pig's stomach. Scrub with coarse salt and corn flour for serveral minutes.

2. Wash pig's stomach and blanch in water for a while with ginger and Chinese onion. Dry it and cut into pieces. Set aside.

3. Put all the ingredients into the pot with 15 bowls of water. Boil for about $2^1/_2$ hours and it is done.

材料：

人參（高麗參或石柱參）5錢、蓮子（去蕊）2両、豬肚1個、扁豆1両

做法：

1. 豬肚剪開，以粗鹽及豆粉洗擦數分鐘；

2. 豬肚用水沖至乾淨後，再以薑葱汆水，瀝乾切件備用；

3. 把全部材料一同放進煲中，加15碗水，煲約2¹/₂小時後，即可飲用。

功效

健脾益氣，補氣益腎，去濕利水。對產後脾腎兩虛、水濕內停、身腫水腫等產婦有幫助。

Strengthen the spleen and benefit Qi, replenish Qi and tonify the kidney, dispel dampness and promote body fluid. It is helpful for delivered women who are infirm in both spleen and kidney, dampness within and bodily swelling.

中醫師提提您

若怕清潔豬肚不便，亦可使用豬䐗。當然利水去腫功效則以豬肚為佳。

If you worry that the pig's stomach is not easy to clean, you can use shin of pork instead. However, the effect of promote body fluid and diminishing swelling will be lessened when using shin of pork.

坐月必食

茶飲

補粥

燉品

補湯

飯麵

當歸豬䐡補腎湯
Soup for Replenish Kidney with Chinese Angelica Root and Shin of Pork
(4人份量 for 4 persons)

適用於：	順產、剖腹產、哺乳、不哺乳
宜：	四季
補：	補益氣血
惡露清：	後

Ingredients:

20g whole Chinese angelica root, 12g steamed and processed ginseng, 1 shin of pork, 12g Chinese wolfberry fruit, 3 slices fresh ginger

Steps:

1. Wash shin of pork and blanch in water for a while. Take out and set aside.

2. Wash all other ingredients separately. Set aside.

3. Put all ingredients into the pot with 12 bowls of water. Boil for about 2$^1/_2$ hours and it is done.

材料：

當歸身5錢、紅參3錢、豬䐡1個、杞子3錢、生薑3片

做法：

1. 豬䐡洗淨，汆水，備用；

2. 其他材料分別洗淨，備用；

3. 全部材料加入煲內，加12碗水，煲約2$^1/_2$小時後，即可飲用。

功效

補氣益血，祛風補腎。對產後血虛、風重頭暈、面色蒼白、神疲怠倦之產婦有幫助。

Replenish the blood and benefit Qi, anti-rheumatics and replenish the kidney. It is helpful for delivered women who have infirmity in blood circulation, with wind syndrome, pale in face, and get tired easily.

中醫師提提您

產婦待惡露乾淨後食用。陰虛內熱及外感表證者不宜飲用。每星期可飲用1至2次。

Delivered women should drink this soup after her lochia is cleared. This soup is not suitable for those who have internal heat due to Yin deficiency or syndrome of external contraction. It can be drunk once or twice per week.

坐月必食

茶飲

補粥

燉品

補湯

飯餸

靈棗烏雞湯
Soup with Ganoderma Lucidum, Nan Dates and Black Chicken
(4人份量 for 4 persons)

適用於：順產、剖腹產、哺乳、不哺乳

宜：四季

補：益氣健腎

惡露清：前、後

Ingredients:

12g red ganoderma lucidum, 10 nan dates, 80g He Shou Wu, 1 black chicken, 3 slices fresh ginger

Seasoning:

a pinch of salt

Steps:

1. Wash the black chicken, cut into pieces and blanch for a while. Then take out and set aside.
2. Wash all other ingredients and put them into the pot with the black chicken and 15 bowls of water.
3. Boil for about 3 hours. Add some salt and it is done.

材料：

赤靈芝3錢、南棗10粒、何首烏2兩、烏雞1隻、生薑3片

調味料：

鹽少許

做法：

1. 烏雞洗淨、切件、汆水備用；
2. 其他材料一同洗淨後，加水15碗與烏雞同下煲中；
3. 煲約3小時，略加鹽調味即可飲用。

功效

強壯補虛、補血祛風、補腎滋陰。對產後精神不振、頭髮脫落、白髮頻生、氣血不足、血虛頭暈之產婦有幫助。

Strengthen the body and supplement vacuity, replenish the blood and anti-rheumatics, replenish the kidney and nourish Yin. It is helpful for delivered women who are listless, losing hair, whitening of hair, insufficient blood and Qi and infirmity in blood and dizziness.

中醫師提提您

四季皆宜，惡露乾淨前後皆可飲用。若買不到烏雞，可選用竹絲雞代替，功效相同。

This soup is suitable for all seasons and can be consumed before or after the lochia is cleared. If black chicken is not available, you can use silky fowl instead, which has the same effectiveness.

坐月必食

茶飲

補粥

燉品

補湯

飯餸

粟米魚肚蛋花湯
Soup with Corn, Fish Maw and Egg
（4人份量 for 4 persons）

適用於：順產、剖腹產、哺乳、不哺乳	
宜：四季	
補：健脾開胃	
惡露清：前、後	

Ingredients:
80g dried fish maw, 1 can of corn, 1 egg, 160g lean pork

Seasoning:
a pinch of salt

Steps:
1. Soak dried fish maw in water, wash and cut into small pieces. Set aside.
2. Beat egg and mix well. Set aside.
3. Put the can of corn and lean pork into the pot with 12 bowls of water and bring to boil over high heat. After boils, cook for another 30 minutes with medium heat.
4. Put the fish maw into the soup. When it boils again, put in the egg and season with some salt. The soup is ready to serve when the egg is done.

材料：
魚肚2両、粟米湯1罐、雞蛋1隻、瘦肉4両

調味料：
鹽少許

做法：
1. 魚肚浸泡洗淨，切碎，備用；
2. 打雞蛋後拂勻，備用；
3. 將粟米湯罐頭及瘦肉加入煲中，加12碗水，以大火煮沸後，轉中火再煲30分鐘；
4. 把魚肚放進湯中，待湯再煮滾後，加入已拂勻的雞蛋及略加鹽調味，待雞蛋熟透後即可飲用。

功效
健脾開胃，強健體質，增肌補髓。適合產後胃納欠佳、肌肉瘦削、面色不華之產婦飲用。
Strengthen the spleen and promote ingestion, strengthen the body, increases muscles and replenish bone marrow. It is helpful for delivered women who have poor ingestion, leanness, and dull faces.

中醫師提提您
四季皆宜，味道鮮美，製作簡單，營養豐富，易消化吸收，最好於產後頭幾天未能進食較補益的湯水前飲用此湯，增加營養吸收。
This soup is suitable for all seasons. It is tasty and easy to make. It is rich in nutrition and easy to digest and absorb. It is better to drink this soup in the first few days after delivery for absorbing more nutrition when the woman cannot consume soups with more tonic function.

黃芪益氣豬手湯

Soup for Benefit Qi with Astragalus Membranaceus and Pig's Front Leg

(4人份量 for 4 persons)

適用於：順產、剖腹產、哺乳、不哺乳
宜：四季
補：補氣增乳
惡露清：前、後

Ingredients:

20g heterophylly falsetarwort root, 20g astragalus membranaceus, 1 pig's front leg, 8 Chinese dates (rid of pits), 8g tangerine peel, 40g Chinese yam

材料：

太子參5錢、黃芪5錢、豬手1隻、紅棗8粒（去核）、陳皮2錢、淮山1両

Steps:

1. Get rid of the hair of pig's front leg and wash it. Blanch for a while and set aside.
2. Wash all other ingredients separately. Set aside.
3. Put all ingredients into a pot with 15 bowls of water. Cook for about 3 hours and it is done.

做法：

1. 豬手去毛洗淨，汆水，備用；
2. 其他材料分別洗淨，備用；
3. 全部材料一同放進煲中，加15碗水，煲約3小時後即可飲用。

功效

有健脾益氣、補血增乳、開胃益食之功。適合產後氣血虛弱、乳汁不足、胃納欠佳、肌肉瘦削之產婦飲用。

Strengthen spleen and benefit Qi, replenish the blood and increase milk supply, enhance and induce appetite. It is helpful for delivered women who are deficient in blood and Qi, not enough milk, poor ingestion, and leanness.

中醫師提提您

四季皆宜，性質平和，若產婦消化功能欠佳者，可改用豬腱棄用豬手，待體質逐漸改善後轉用豬手。

This soup is suitable for all seasons and mild in nature. If delivered woman has poor digestion after delivery, use shin of pork instead of pig's front leg until her body gets strengthened.

坐月必食　茶飲　補粥　燉品　補湯　飯麵

紫河車蓮藕牛腒湯
Soup with Dried Placenta, Lotus Root and Beef

(4人份量 for 4 persons)

適用於：順產、剖腹產、哺乳、不哺乳
宜：四季
補：補腎養血
惡露清：後

Ingredients:
1/2 dried placenta, 640g lotus root, 3 slices fresh ginger, 8 Chinese dates (rid of pits), 12g Chinese wolfberry fruit, 320g shin of beef

Seasoning:
a pinch of salt

Steps:
1. Wash and slice the shin of beef, set aside.
2. Soak the dried placenta in water and let it swollen. Wash the lotus root and cut into pieces. Set aside.
3. Put the dried placenta, lotus root and other ingredients into the pot with 15 bowls of water. Cook for about 2 1/2 hours.
4. Put the shin of beef in. After the shin of beef is done, season with some salt and it is done.

材料：
紫河車半個、蓮藕1斤、生薑3片、紅棗8枚（去核）、杞子3錢、牛腒半斤

調味料：
鹽少許

做法：
1. 牛腒洗淨，切片，備用；
2. 紫河車洗淨浸泡，蓮藕洗擦，切件，備用；
3. 紫河車、蓮藕與其他材料一同放進煲中，加15碗水，煲約2 1/2小時；
4. 加入牛腒，待牛腒熟透後，略加鹽調味即可飲用。

功效

益氣養血，強筋健體。適合產後氣血虧損、面色無華、腰酸尿頻、貧血、神疲怠倦之產婦飲用。
Benefit Qi and nourish blood, strengthen the muscles and body. It is helpful for delivered women who have deficiency of blood and Qi, dull faces, back pain and frequent urination, anaemia, and tiredness.

中醫師提提您

待惡露乾淨後作補虛強壯湯飲用，每星期可作2次飲用。
This soup can be served as a tonic drink for infirmity and strengthening the body after the lochia is cleared. It can be consumed twice a week.

營養進補坐月食譜 115

坐月必食

茶飲

補粥

燉品

補湯

飯餸

海參核桃鴿肉湯
Soup with Sea Cucumber, Walnut, and Pigeon

（4人份量 for 4 persons）

適用於：順產、剖腹產、哺乳、不哺乳	
宜：四季	
補：補腎益腎	
惡露清：前、後	

Ingredients:

120g sea cucumber, 80g walnut (shelled), 8g dried longon, 8 Chinese dates (rid of pits), 1 pigeon, 3 slices fresh ginger, 8g tangerine peel

材料：

海參3両、核桃2両（去殼）、圓肉2錢、紅棗8枚（去核）、白鴿1隻、生薑3片、陳皮2錢

Seasoning:

a pinch of salt

調味料：

鹽少許

Steps:

1. Soak the sea cucumbers in water and let it swollen; wash and cut into pieces, set aside.
2. Get rid of the organs of the pigeon and wash; blanch for a while. Soak the tangerine peel in water and get rid of the white stuff of the peel, set aside.
3. Put all ingredients except the sea cucumbers into the pot with 15 bowls of water. Cook for about 1½ hours.
4. Put the sea cucumbers in and cook for about 45 minutes. Add some salt and it is done. Consume both soup and meat.

做法：

1. 海參浸發後洗淨，切件，備用；
2. 白鴿去內臟洗淨，汆水；陳皮浸泡後，去囊，備用；
3. 核桃、圓肉、紅棗、生薑、陳皮及白鴿一同放進煲中，加15碗水，煲約1½小時；
4. 放入海參後再煲約45分鐘，加鹽調味後即可飲湯食肉。

功效

益腎健脾，填精生髓，強壯體質。適合產後出現腰酸腿軟、神疲怠倦、氣血不足、肌肉瘦削之產婦飲用。

Tonify the kidney and strengthen the spleen, support essence and generate bone marrow, strengthen the body. It is helpful for delivered women who have back pain and weak legs, tiredness, deficiency of Qi and blood, and leanness.

中醫師提提您

惡露乾淨前後均可飲用。常服此湯有強身補腦之功效，男女老幼皆宜。

This soup can be consumed before and after lochia is cleared. Drinking this soup regularly can strengthen your body and replenish the brain. It is suitable for all ages and gender.

坐月必食

茶飲

補粥

燉品

補湯

飯餸

海參淮杞豬骨湯

Soup with Sea Cucumber, Chinese Yam,
Chinese Wolfberry Fruits and Pig's Bones

（4人份量 for 4 persons）

適用於：	順產、剖腹產、哺乳、不哺乳
宜：	四季
補：	滋陰補腎
惡露清：	後

Ingredients:

120g sea cucumber, 40g Chinese yam, 40g Chinese wolfberry fruit, 5 nan dates, 640g pig's bones

Seasoning:

a pinch of salt

Steps:

1. Wash the sea cucumbers, cut into pieces and set aside. Wash the pig's bones and blanch for a while.

2. Put all ingredients except the sea cucumbers into the pot with 15 bowls of water. Cook for about 1 hour.

3. Put the sea cucumbers in and cook for another 45 minutes. Add some salt and it is done. Consume both soup and meat.

材料：

海參3両、淮山1両、杞子1両、南棗5枚、豬骨1斤

調味料：

鹽少許

做法：

1. 海參洗淨，切件備用；豬骨洗淨，汆水；

2. 把藥材及豬骨一同放進煲中，加15碗水，煲約1小時；

3. 放入海參再煲約45分鐘，加鹽調味後即可飲湯食肉。

功效

養血生髓，健脾開胃，滋補強壯。適合產後腰酸腿軟、神疲、氣血不足之產婦飲用。

Nourish the blood and generate bone marrow, strengthen the spleen and induce appetite, moisten and strengthen the body. It is helpful for delivered women who have back pain and weak legs, tiredness, and deficiency of Qi and blood.

中醫師提提您

四季皆宜、惡露乾淨後飲用。外感表證患者不宜服用。

This soup can be consumed in all seasons but after the lochia is cleared. It is not suitable for people who have syndromes of external contraction.

坐月必食

茶飲

補粥

燉品

補湯

飯餸

補湯篇

巴戟溫腎益肝湯

Soup for Warming the Kidney and Replenishing the Liver with Morinda Root

(4人份量 for 4 persons)

適用於：順產、剖腹產、哺乳、不哺乳
宜：四季
補：滋補肝腎
惡露清：後

Ingredients:

16g morinda root, 20g Chinese wolfberry fruit, 120g pig's liver, 3 slices fresh ginger, 2 candied jujubes

Seasoning:

a pinch of salt

Steps:

1. Wash the pig's liver, cut into pieces and set aside.

2. Wash all other ingredients and put into the pot with 12 bowls of water. Cook for about 1½ hours.

3. Put the pig's liver in and cook for about 10 minutes. Add some salt and it is done. Consume both soup and meat.

材料：

巴戟4錢、杞子5錢、豬肝3両、生薑3片、蜜棗2枚

調味料：

鹽少許

做法：

1. 沖洗豬肝的血水後，切件備用；

2. 其他材料洗淨後，放進煲中，加水12碗，煲約1½小時；

3. 放進豬肝，待10分鐘後，略加鹽調味，即可飲湯吃肉。

功效

補腎強精，益肝養血。適合產後體虛、肝腎不足、筋骨酸軟、尿頻耳鳴之產婦飲用。

Replenish the kidney and strengthen the essence, replenish the liver and blood. It is helpful for delivered women who have an infirm body, insufficient function of liver and kidney, pain in muscles and bones, and frequent urination and tingling of the ears.

中醫師提提您

四季皆宜，惡露乾淨後飲用。陰虛火旺及外感表證患者忌服。

This soup can be consumed in all seasons but only after the lochia is cleared. It is not suitable for people who have flaring of fire due to Yin deficiency and syndromes of external contraction.

坐月必食

茶飲

補粥

燉品

補湯

飯餸

木瓜花生通乳湯
Soup for Inducing Milk with Papaya and Peanuts
（4人份量 for 4 persons）

適用於：	順產、剖腹產、哺乳
宜：	四季
補：	健脾增乳
惡露清：	前、後

Ingredients:
160g peanuts, 12g rice paper plant pith, 10 Chinese dates (rid of pits), 1 papaya, 20g Chinese yam, 8 chicken feet

材料：
花生4両、通草3錢、紅棗10枚（去核）、木瓜1個、准山5錢、雞腳8隻

Seasoning:
a pinch of salt

調味料：
鹽少許

Steps:
1. Peel off the skin of the papaya and get rid of the seeds, wash and cut into blocks, set aside.
2. Wash the chicken feet; blanch for a while and set aside.
3. Wash other ingredients separately. Set aside.
4. Put all ingredients into the pot with 12 bowls of water. Cook for about 2 hours. Season with some salt and it is done.

做法：
1. 木瓜去皮及籽，洗淨後切件備用；
2. 雞腳洗淨、汆水備用；
3. 其他材料洗淨，備用；
4. 將全部材料放入煲中，加12碗水，煲約2小時，略加鹽調味即可飲用。

功效
補腎強骨、健脾益胃、滋補氣血、增乳通乳汁。適合產後氣血兩虛、乳汁不充、肌肉瘦削、面色萎黃之產婦飲用。
Replenish the kidney and strengthen the bones, strengthen the spleen and nourish the stomach, nourish and replenish Qi and blood, increases and induces milk. It is helpful for delivered women who are deficient in Qi and blood, insufficient in milk, leanness, and yellowing of faces.

中醫師提提您
四季皆宜，每星期則可飲用2至3次，功效更佳。
This soup can be consumed in all seasons. It is even better if consumed 2 to 3 times a week.

坐月必食

茶飲

補粥

燉品

補湯

飯餸

補湯篇

百年好合清雞湯
Chicken Soup with Dried Lily
(4人份量 for 4 persons)

適用於：順產、剖腹產、哺乳、不哺乳	
宜：四季	
補：安神補虛	
惡露清：前、後	

Ingredients:
80g lotus seeds (without pith), 80g lily bulb, 8g tangerine peel, 80g Fu Shen, 8 Chinese dates (rid of pits), 1 chicken

Seasoning:
a pinch of salt

Steps:
1. Wash the chicken and get rid of the fats and skin; cut into pieces and blanch for a while, set aside.
2. Wash all other ingredients separately. Set aside.
3. Put all ingredients into the pot with the chicken and 15 bowls of water. Cook for about 2 hours. Season with some salt and it is done.

材料：
蓮子（去蕊）2両、百合2両、陳皮2錢、茯神2両、紅棗8枚（去核）、雞1隻

調味料：
鹽少許

做法：
1. 光雞洗淨後去皮脂，切件、汆水備用；
2. 其他材料分別洗淨，備用；
3. 將全部材料一同放進煲中，加15碗水，煲約2小時，略加鹽調味即可飲用。

功效
開胃健脾、寧心安神、潤肺益氣。適合產後血虛、失眠不寐、心悸多夢、神疲怠倦之產婦飲用。
Enhances appetite and strengthen spleen, clams the mind, moistens the lung and benefit Qi. It is helpful for delivered women who have deficiency of blood, insomnia, palpitation and dreaming a lot, and tiredness.

中醫師提提您
四季皆宜、性質平和，每星期可作1至2次。惡露乾淨前後均可飲用。
This soup can be consumed in all seasons and mild in nature. You can drink it once or twice a week, and before or after lochia is cleared.

當歸補血竹絲雞湯
Soup for Replenish Blood with Chinese Angelica Root and Silky Fowl
（4人份量 for 4 persons）

適用於：順產、剖腹產、哺乳、不哺乳	
宜：冬季	
補：補血祛風	
惡露清：後	

Ingredients:
20g Chinese angelica root, 8g hemlock parsley, 2 slices fresh ginger, 8 Chinese dates (rid of pits), 16g Chinese yam, 40g dang shen, 1 silky fowl

材料：
當歸5錢、川芎2錢、生薑2片、紅棗8枚（去核）、淮山4錢、黨參1兩、竹絲雞1隻

Seasoning:
a pinch of salt

調味料：
鹽少許

Steps:
1. Wash the chicken and get rid of the fats and skin. Cut into pieces and blanch for a while. Set aside.
2. Wash all other ingredients separately and set aside.
3. Put them into the pot with the chicken and 15 bowls of water. Cook for about 2½ hours. Season with some salt and it is done.

做法：
1. 竹絲雞洗淨後，去皮脂，切件、汆水備用；
2. 其他材料分別洗淨，備用；
3. 將全部材料一同放進煲中，加15碗水，煲約2½小時，略加鹽調味即可飲用。

功效
補血養血，健脾益氣，強壯體質。適合產後血虛、出現頭目眩暈頭痛、手足不溫等產婦飲用。
Replenish and nourish the blood, strengthen spleen and benefit Qi, strengthens the body. It is helpful for delivered women who have deficiency of blood, dizziness, headache, and cold in hands and feet.

中醫師提提您
冬季較佳，性質偏溫、寒性體質人士飲用效果更好。宜於惡露乾淨後飲用。 脾濕內蘊、陰虛火旺及外感表證患者不宜飲用。
This soup is better consumed in winter because it is "warm" in nature and good for people with "internal cold" condition. You should drink this soup only after lochia is cleared. People with syndrome of damp stagnating in spleen, having flaring of fire due to Yin deficiency or syndromes of external contraxtion should not consume this soup.

坐月必食

茶飲

補粥

燉品

補湯

飯餸

充乳湯
Soup for Increasing Milk
（4人份量 for 4 persons）

適用於：順產、剖腹產、哺乳
宜：四季
補：健脾增乳
惡露清：後

Ingredients:

20g rice paper plant pith, 20g Chinese wolfberry fruit, 40g Chinese yam, 20g cowherb seed, 320g shin of pork

Steps:

1. Wash the shin of pork and blanch in boiling water for a while. Set aside.
2. Wash all other ingredients separately. Set aside.
3. Put all ingredients into a pot with 12 bowls of water. Cook for about 2 hours and it is done.

材料：

通草5錢、杞子5錢、淮山1両、王不留行5錢、豬䐑半斤

做法：

1. 豬䐑洗淨、汆水備用；
2. 其他材料分別洗淨，備用；
3. 將全部材料一同放進煲中，加12碗水，煲約2小時，即可飲用。

功效

通乳健脾、滋陰補腎。適合產後氣血不足、乳汁缺乏的產婦，有助補充營養、增加乳汁及乳汁分泌。

Induce milk and strengthen spleen, moisten Yin and replenish kidney. It is helpful for delivered women who are deficient in Qi and blood, and lack of milk. It can supply nutrition and increase milk supply for the delivered women.

中醫師提提您

四季皆宜，待惡露乾淨後飲用更佳。每星期可服2次。

This soup can be consumed in all seasons. It is better to drink after lochia is cleared and can be drunk twice a week.

歸芪大棗牛肉湯
Soup with Chinese Angelica Root, Astragalus Membranaceus, Jujubes and Beef
（4人份量 for 4 persons）

適用於：	順產、剖腹產、哺乳、不哺乳
宜：	冬季
補：	補中益氣
惡露清：	後

Ingredients:
12g whole Chinese angelica root, 40g astragalus membranaceus, 8 jujubes, 3 slices fresh ginger, 160g beef, 12g tangerine peel

Seasoning:
a pinch of salt

Steps:
1. Wash the beef. Slice and set aside.
2. Wash and slice the ginger. Wash all other ingredients and put them into the pot with 12 bowls of water. Cook for about 1½ hours.
3. Put the beef in and cook for another ½ hour. Season with some salt and it is done.

材料：
全當歸3錢、北芪1両、大棗8粒、生薑3片、牛肉4両、陳皮3錢

調味料：
鹽少許

做法：
1. 牛肉洗淨後，切片備用；
2. 生薑洗淨後切片；其他材料洗淨後放進煲中，加12碗水，煲約1½小時；
3. 放入牛肉後，再煲½小時，略加鹽調味即可飲用。

功效
補益養血，強壯體質，祛風健胃。對產後氣血兩虛、頭暈不寐、心悸失眠、自汗不適之產婦適用。
Replenish and nourish the blood, strengthens the body, anti-rheumatics and strengthens stomach. It is helpful for delivered women who are infirm in Qi and blood, having dizziness and insomnia, and palpitation and sweat a lot.

中醫師提提您
四季皆宜，可於惡露乾淨後飲用。
This soup can be consumed in all seasons. You can drink it after lochia is cleared.

坐月必食

茶飲

補粥

燉品

補湯

飯餸

烏豆參芪豬䐹湯

Soup with Black Bean, Dang Shen, Astragalus Membranaceus and Shin of Pork

（4人份量 for 4 persons）

適用於：	順產、剖腹產、哺乳、不哺乳
宜：	冬季
補：	補腎烏髮
惡露清：	後

Ingredients:

80g black beans, 40g dang shen, 20g astragalus membranaceus, 320g shin of pork, 8 Chinese dates (rid of pits)

Steps:

1. Wash the black beans. Fry them in the pan for a while and set aside.

2. Wash shin of pork and blanch it in boiling water for a while. Set aside.

3. Wash all other ingredients and put them into the pot with black beans and shin of pork with 12 bowls of water. Cook for about $2^1/_2$ hours and it is done. Consume both meat and soup.

材料：

烏豆2両、黨參1両、北芪5錢、豬䐹半斤、紅棗8枚（去核）

做法：

1. 烏豆洗淨後，落鑊略炒，備用；

2. 豬䐹洗淨、汆水備用；

3. 其他材料分別洗淨，與烏豆、豬䐹一同放進煲中，加12碗水，煲約2½小時後即可飲湯吃肉。

功效

補血益腎，健脾提氣。對產後氣血不足、面色萎黃、白髮頻生、頭髮稀疏、神疲怠倦之產婦有幫助。

Replenish the blood and kidney; strengthen spleen and benefit Qi. It is helpful for delivered women who have deficiency of Qi and blood, yellowing of faces, having white hair or thin hair, and tiredness.

中醫師提提您

惡露乾淨後飲用。若想增強補益力，豬䐹可改用烏雞，但要注意藥性較溫熱，陰虛內熱者不宜。

This soup can be consumed after lochia is cleared. If you would strengthen the effect, use black chicken instead of shin of pork. However, it will become "warm and hot", which is not suitable for those who have internal heat due to Yin deficiency.

金針木耳排骨湯

Soup with Hemerocallis Flava, Wood Fungi and Rib

(4人份量 for 4 persons)

適用於：順產、剖腹產、哺乳、不哺乳
宜：四季
補：疏肝去瘀
惡露清：前

Ingredients:

40g hemerocallis flava, 40g wood fungus, 10 Chinese dates (rid of pits), 40g dang shen, 640g ribs

Steps:

1. Wash hemerocallis flava and wood fungus after soaking them in water. Drain and set aside.
2. Wash the ribs and blanch in boiling water for a while. Set aside.
3. Wash all other ingredients and put them into the pot with hemerocallis flava, wood fungus, ribs, and 12 bowls of water. Cook for about 2 hours and it is done.

材料：

金針1両、白背木耳1両、紅棗10枚（去核）、黨參1両、排骨頭1斤

做法：

1. 金針、木耳浸發後洗淨，瀝乾備用；
2. 排骨頭洗淨、汆水備用；
3. 其他材料分別洗淨，與金針、木耳及排骨一同放進煲中，加12碗水，煲約2小時後即可飲用。

 功效

活血去瘀，補氣益血，增強體力，疏肝理氣，鬆弛神經。對產後惡露不絕、精神不振、氣血兩虛、心煩少寐、肝鬱氣結之產婦有幫助。
Promote blood circulation and remove blood stasis, replenish Qi and nourish blood, strengthen the body, dispels the stagnated liver-qi and clams the mind. It is helpful for delivered women who have too much lochia, listless, deficiency of Qi and blood, unstabled mind and troubled sleep, liver - qi stagnation.

 中醫師提提您

四季皆宜，宜於惡露排出的日子服用，有助去瘀生新血。
This soup can be consumed in all seasons and is best when there is lochia. It helps relieving blood stasis and creating new blood.

坐月必食

茶飲

補粥

燉品

補湯

飯餸

補腎烏髮豬骨湯
Soup for Replenish Kidney and Blackening Hair with Pig's Bones
(4人份量 for 4 persons)

適用於：順產、剖腹產、哺乳、
　　　　不哺乳

宜：四季

補：補血益腎

惡露清：前、後

Ingredients:

8g hair moss, 16g Chinese wolfberry fruit, 20g Shou Wu, 80g black beans, 2 slices fresh ginger, 640g pig's bones, 3 candied jujubes

Steps:

1. Soak hair moss in water until it is swollen. Drain and set aside.
2. Wash all other ingredients and put them into the pot with 15 bowls of water. Cook for about 2$^1/_2$ hours and it is done.

材料：

髮菜2錢、杞子4錢、首烏5錢、烏豆2両、生薑2片、豬骨1斤、蜜棗3粒

做法：

1. 髮菜浸發，洗淨，瀝乾備用；
2. 其他材料分別洗淨，把全部材料一同放進煲中，加水15碗，煲約2$^1/_2$小時即可飲用。

 功效

補血滋腎、美容烏髮。對產後肝腎虛損、氣虛神疲、腰膝酸軟、尿頻白髮、灰黑面斑者有幫助。

Replenish blood and nourish kidney, improve the appearance and blacken hair. It is helpful for delivered women who are infirm in liver and kidney, deficient in Qi and with tiredness, aching and limp at the back and the knees, frequent urination, whitening of hair, and black or gray speckles in faces.

 中醫師提提您

四季皆宜，惡露乾淨前後均可飲用。一家大細、男女老幼皆宜。

This soup can be consumed in all seasons, and before and after lochia is cleared. It is suitable for all family members and both men and women.

坐月必食

茶飲

補粥

燉品

補湯

飯餸

參芪魚頭豆腐湯

Soup with Dang Shen, Astragalus Membranaceus Fish Head and Bean Curd

(4人份量 for 4 persons)

適用於：順產、剖腹產、
　　　　哺乳、不哺乳

宜：四季

補：健胃祛風

惡露清：前

Ingredients:
40g dang shen, 20g astragalus membranaceus, 1 big fish head, 2 blocks bean curd, 3 slices fresh ginger, 3 blocks Chinese onions, 8 Chinese dates (rid of pits)

Seasoning:
a pinch of salt

Steps:
1. Wash all herbs, ginger and Chinese onions. Set aside.
2. Wash the big fish head and bean curds separately. Drain and set aside.
3. Put all the herbs and bean curds into the pot with 12 bowls of water. Cook for about 1½ hours.
4. Put the big fish head in and boil until the fish head is done. Season with some salt and it is done.

材料：
黨參1両、北芪5錢、大魚頭1個、豆腐2磚、生薑3片、葱3段、紅棗8枚（去核）

調味料：
鹽少許

做法：
1. 所有藥材、薑、葱洗淨，備用；
2. 魚頭、豆腐分別洗淨，瀝乾水份，備用；
3. 先把藥材及豆腐一同放進煲中，加12碗水，煲約1½小時；
4. 放入魚頭，煲至魚頭熟透後，略加鹽調味即可飲湯吃肉。

功效

健脾益氣，補血養血，祛風健胃。適合產後氣血兩虛、頭暈目眩、胃納欠佳、神疲怠倦之產婦飲用。
Strengthen spleen and benefit Qi, replenish and nourish the blood, anti-rheumatics and strengthen the stomach. It is helpful for delivered women who have deficiency of Qi and blood, dizziness, poor ingestion and tiredness.

中醫師提提您

此湯可於惡露乾淨前後飲用，有利產後子宮收縮，促進惡露排出。
This soup can be consumed before and after lochia is cleared. It helps uterus to contract so that more lochia can be discharged.

坐月必食

茶飲

補粥

燉品

補湯

飯餸

蟲草固本培元酒
Wine for Strengthening the Body with Chinese Cordyceps
(1人份量 for 1 person)

適用於：順產、剖腹產、哺乳、不哺乳
宜：四季
補：固本培元
惡露清：後

Ingredients:
20g Chinese cordyceps, 80g steamed and processed ginseng, 40g wolfberry fruit, 80g astragalus membranaceus, 80g Shou Wu, 20 nan dates, 12g tangerine peel, 40g rhizoma polygonati, 20g privet fruit, 3.2kg rice wine

Steps:
1. Wash all ingredients with rice wine. Then put into a wine bottle to be sealed.
2. Put in the wine and seal it for 2 weeks.
3. Drink 15ml per day.

材料：
冬蟲夏草5錢、紅參2両、枸杞子1両、北芪2両、首烏2両、南棗20粒、陳皮3錢、黃精1両、女貞子5錢、米酒5斤

做法：
1. 全部材料以米酒沖洗後，一同放進存氣的酒樽內；
2. 加進米酒5斤，封蓋後待2週後飲用；
3. 每日一次，每次15毫升。

功效
扶正固本，調理陰陽；益氣養血，滋補肝腎。對產後血虛、元氣虛損、肝腎不足、食少怠倦之產婦有幫助。
Strengthen the body resistance, regulate Yin and Yang, benefit Qi and nourish blood, moisten and nourish liver and kidney. It is helpful for delivered women who have deficiency of blood, losing Qi, insufficiency of liver and kidney, poor appetite and tiredness.

中醫師提提您
可於惡露乾淨後開始飲用。男女皆宜，四季均可。成人或產婦之日常補酒，一樽可飲用3個月至半年。
This can be consumed after lochia is cleared. It is suitable for both men and women, and at all seasons. Adult and delivered women can treat this wine as daily tonic wine. 1 bottle of wine can be drunk about 3 to 6 months.

第三部份：

坐月健美
營養飯餸

（此部分由胡美怡註冊營養師提供）

營養飽菜篇

營養成份 Nutrition information

碳水化合物(Carbohydrate)	☑	脂肪(Fat)	☑
		低脂(Low Fat)	
蛋白質(Protein)	☑	礦物質(Minerals)	☑
		鈣質(Calcium)	
維他命(Vitamins)	☐	纖維(Dietary Fibre)	☐

白汁蘑菇雞柳焗飯

Baked Chicken Rice with Mushroom in Cream Sauce

(2人份量 for 2 persons)

Ingredients:
60g boiled chicken meat (diced), 40g sliced mushroom, 1 1/2 bowl rice, 1/4 canned chicken broth, 2 pieces low-fat cheese (40g), suitable amount of shredded Parmesan cheese, 30g onion, 15g chopped garlic, 4 tbsp low-fat milk powder

Utensil:
Oven

Steps:
1. Preheat oven to 200 °C.

2. Heat oil in pan. Stir-fry garlic and onion until fragrant. Add in sliced mushroom, chicken dices and rice, and stir-fry.

3. Blend chicken broth and low-fat milk powder together and spread it over the stir-fried rice. Cook until sauce is reduced. Add in low-fat cheese and cook until cheese melts.

4. Pour (3) into a baking tray. Sprinkle with suitable amount of shredded cheese. Bake for 3 minutes until cheese melts and browns.

材料：
白灼雞肉（切丁）60克、蘑菇（切片）40克、飯1碗半、罐頭雞湯1/4罐、低脂芝士2塊（40克）、巴馬乾芝士絲適量、洋葱30克、蒜茸15克、低脂奶粉4湯匙

工具：
焗爐

做法：
1. 預熱焗爐至攝氏200度；

2. 熱油鍋，加入少許蒜茸、洋葱爆香後，再加入蘑菇片、雞丁及飯一起拌炒；

3. 雞湯加入低脂奶粉拌勻後，注入炒飯中，煮至收汁，再加入低脂芝士煮至融化；

4. 將（3）的飯放入烤盤中，灑上適量芝士絲，放入焗爐焗約3分鐘，直至芝士絲融化上色即可。

營養師提提您 Dietitian's Tips

低脂奶及低脂芝士提供豐富鈣質，脂肪量又低。餵哺母乳的媽媽可以不用洋葱、蒜頭。

Milk and cheese is rich in calcium and low-fat. Breast-feeding mothers can leave out onion and garlic in the recipe.

營養成份 Nutrition information

碳水化合物(Carbohydrate)	☑	脂肪(Fat)		☑
		奧米加3脂肪酸(Omega-3 fatty acids)：三文魚(salmon)		
蛋白質(Protein)	☑	礦物質(Minerals)		☑
		磷質(Phosphorous)		
維他命(Vitamins)	☑	纖維(Dietary Fibre)		☑
葉酸(Folic acid)：蘆筍(asparagus)				
維他命A及D(Vit. A & D)：蛋(egg)				

三文魚蘆笋炒飯
Fried Rice with Salmon and Asparagus
（2人份量 for 2 persons）

Ingredients:
1 piece salmon (120g), 100g asparagus, 1 tbsp grated ginger, 60g chopped Chinese celery, 1 stalk Chinese onion (chopped), 1 egg, 1 bowl rice

Seasoning:
small amount of pepper, $\frac{1}{2}$ tsp fish sauce

Steps:
1. Steam salmon. Debone and cut into small pieces. Set aside.
2. Rinse asparagus. Scald until done. Drain dry and cut into cubes. Set aside.
3. Beat egg in a bowl to form egg sauce. Set aside.
4. Heat 1 tbsp of oil in a pan. Sauté Chinese onion, Chinese celery and ginger until fragrant. Add in asparagus, rice and seasoning. Stir-fry until rice begins to pop.
5. Add in salmon and egg sauce. Keep stir-frying until the egg is done.

材料：
三文魚1塊（120克）、蘆笋100克、薑茸1湯匙、芹菜粒60克、葱1棵（切粒）、雞蛋1個、白飯1碗

調味料：
胡椒粉少許、魚露 $\frac{1}{2}$ 茶匙

做法：
1. 三文魚蒸熟後拆肉，切成小塊，備用；
2. 蘆笋洗淨，放入滾水中燙熟，取出瀝乾水份，切粒，備用；
3. 雞蛋打入碗中打成蛋汁，備用；
4. 燒熱1湯匙油，放入葱粒、芹菜粒及薑茸炒香，加入蘆笋、白飯及調味料炒勻至飯散開；
5. 加入三文魚及蛋汁，炒至蛋熟即可。

坐月必食
茶飲
補粥
燉品
補湯
飯餸

 營養師提提您 Dietitian's Tips

三文魚含有奧米加3脂肪酸，有助抗產後抑鬱。哺乳媽媽食三文魚，母乳內也含有奧米加3脂肪酸，有助嬰兒視力及腦部發展。
Salmon contains omega-3 fatty acids which help to prevent postpartum depression. For breast-feeding mothers who eat salmon, her breast milk also contains Omega-3 fatty acids, which are good for baby's eyesight and brain development.

營養餸菜篇

營養成份 Nutrition information

碳水化合物(Carbohydrate)	☑	脂肪(Fat)	☑
		低脂(Low Fat)	
蛋白質(Protein)	☑	礦物質(Minerals)	☐
維他命(Vitamins)	☐	纖維(Dietary Fibre)	☐

黃鱔杞子焗飯

Clay-pot Rice with Finless Eel and Chinese Wolfberry Fruit

（2人份量 for 2 persons）

Ingredients:

1¹/₂ bowl rice, 1 finless eel (around 120g), 8g Chinese wolfberry fruit, suitable amount of chopped ginger and chopped Chinese onion, small amount of soy sauce

Steps:

1. Rub finless eel with salt and corn flour to remove sticky stuff. Cut into sections. Set aside.

2. Put rice and suitable amount of water in a clay pot. Bring to boil.

3. When rice begins to boil, add in finless eel and Chinese wolfberry fruit. Cover and cook until done.

4. When rice is cooked, add in ginger, Chinese onion and a touch of soy sauce. Mix well and serve.

材料：

白飯1碗半、黃鱔1條（約120克）、杞子2錢、薑（切碎）適量、葱（切粒）適量、豉油少許

做法：

1. 利用鹽和生粉刷去黃鱔上的潺，切段備用；

2. 米和適量水放入瓦煲內煮飯；

3. 飯滾即放入黃鱔和杞子，蓋上煲蓋焗至熟；

4. 待飯焗好後才加入薑葱和少許豉油，拌勻後即可食用。

營養師提提您 Dietitian's Tips

黃鱔含有豐富蛋白質和低含量脂肪，產婦食用既補身又不會影響瘦身。

Finless eel is rich in protein and low in fat. It is particularly suitable for the postpartum diet.

坐月必食

茶飲

補粥

燉品

補湯

飯餸

營養成份 Nutrition information

碳水化合物(Carbohydrate)	☑	脂肪(Fat)	☐
蛋白質(Protein)	☑	礦物質(Minerals)	☑
		鈣質(Calcium)	
維他命(Vitamins)	☑	纖維(Dietary Fibre)	☑
維他命A(Vit. A)			
胡蘿蔔素(Carotene)			

焗鮮茄吞拿魚螺絲粉
Baked Fusilli with Tuna
（2人份量 for 2 persons）

坐月必食

茶飲

補粥

燉品

補湯

飯餸

Ingredients:
150g tuna, 2 tomatoes (chopped),150g fusilli, 2 tbsp grated garlic, 1/2 onion (chopped), 1-2 slices low-fat Cheddar cheese, 1 tbsp Parmesan cheese, small amount of olive oil, whole kernel corn (amount as desired)

Utensil:
Oven

Steps:
1. Preheat oven to 20 °C.
2. Boil 1 litre of water with 1 tsp of salt. When water is boiled, change to medium heat and add in fusilli until cooked (about 12 minutes). Drain dry. Add in and mix well with olive oil while still hot. Put fusilli in the baking tray. Set aside.
3. Drain tuna, and then mash it with a fork.
4. Heat olive oil in a pan. Sauté garlic and onion until fragrant. Add in tuna and stir-fry. Add in tomato dices and continue to stir-fry thoroughly. Season with a pinch of sugar, salt and pepper.
5. Pour (4) onto fusilli. Spread Cheddar cheese slices on top and then sprinkle with Parmesan cheese. Bake in oven for about 30 minutes and it is done.

材料：
吞拿魚150克、番茄2個（切碎）、螺絲粉150克、蒜茸2湯匙、洋葱半個（切碎）、低脂車打芝士片1-2片、巴馬乾芝士1湯匙、橄欖油少許、熟粟米粒隨意

工具：
焗爐

做法：
1. 預熱焗爐至攝氏200度；
2. 煮滾1公升水後加入1茶匙鹽，轉中慢火煮熟螺絲粉（約12分鐘），瀝乾水份，趁熱加入橄欖油拌勻，放在焗盤中備用；
3. 吞拿魚瀝乾後用叉攪碎，備用；
4. 燒熱橄欖油，下蒜茸及洋葱粒爆香後，下吞拿魚炒勻，再加入茄粒炒勻，加入少許糖、鹽及胡椒粉調味；
5. 把（4）鋪在螺絲粉上，再鋪上車打芝士，撒下巴馬乾芝士，放入焗爐焗約30分鐘即成。

營養師提提您 Dietitian's Tips
顏色鮮豔，進食容易，且含豐富鈣質，有助預防骨質疏鬆症。
This baked fusilli is colourful and appetitive. It is rich in calcium which help prevent osteoporosis.

營養餐菜篇

營養成份 Nutrition information

碳水化合物(Carbohydrate)	☑	脂肪(Fat)	☑
		低脂(Low Fat)	
蛋白質(Protein)	☑	礦物質(Minerals)	☑
		鐵質(Iron)	
維他命(Vitamins)	☐	纖維(Dietary Fibre)	☐

白汁牛柳
Beef in White Sauce
(2人份量 for 2 persons)

坐月必食

茶飲

補粥

燉品

補湯

飯餸

Ingredients:
200g beef, 2 potatoes, 1 slice old ginger, suitable amount of pea and sliced carrot

Ingredients for white sauce:
1-2 tbsp flour, 1 cup low-fat milk (250ml), 4-5 mushrooms (sliced), small amount of butter

Seasonings:
1/2 tsp salt

Steps:

1. Rinse beef and cut into cubes. Blanch in hot water for 1 minute. Take them out and set aside.

2. Peel potatoes and cut into cubes. Rinse them and set aside.

3. Melt some butter in a pan. Add in flour and cook until paste-like. Pour low-fat milk into the paste and let boil. Add in mushroom slices and stir well. Cook for 1-2 more minutes to form white sauce.

4. Heat oil in a pan. Sauté Chinese onion and ginger until fragrant. Add in cubes of beef and stir-fry for 1-2 minutes. Pour in seasoning and potato cubes. Turn to low heat and cook for 20-30 minutes. Pour white sauce over and serve.

材料：
牛肉200克、馬鈴薯2個、老薑1片、適量青豆、紅蘿蔔片

白汁材料：
1-2湯匙麵粉、低脂奶1杯（250ml）、蘑菇（切片）4-5朵、牛油少許

調味料：
鹽 1/2 茶匙

做法：

1. 牛肉洗淨，切方塊，用沸水先燙煮1分鐘，取出備用；

2. 馬鈴薯去皮切方塊，洗淨備用；

3. 將少許牛油融於鍋中，加麵粉煮成糊狀，注入低脂奶煮滾，放入蘑菇片拌勻，再煮1-2分鐘後即成白汁；

4. 燒熱油鍋，炒香葱薑，再加入牛肉翻炒1-2分鐘後，加調味料及馬鈴薯，用小火煮20-30分鐘，加入白汁拌勻即可。

營養師提提您 Dietitian's Tips

白汁牛肉含有豐富的蛋白質和鐵質，有助預防貧血，非常適合產後食用。
Beef in white sauce is rich in protein and iron, which help prevent anemia. It is particularly suitable for the postpartum diet.

營養成份 Nutrition information

碳水化合物(Carbohydrate)	☐	脂肪(Fat)	☐
蛋白質(Protein)	☑	礦物質(Minerals)	☑
		鐵質(Iron)	
		鋅(Zinc)	
		鎂(Magnesium)	
維他命(Vitamins)	☐	纖維(Dietary Fibre)	☐

紅酒汁燴羊扒
Lamb Chops in Red Wine Sauce
(2人份量 for 2 persons)

Ingredients:
2 pieces lamb chops, 1 tomato, 1/2 onion, 2 cloves garlic, 1/4 tsp salt, 1/4 tsp black pepper

Sauce:
1/2 tbsp oil, 1/4 tsp sugar, 1/2 cup red wine (125ml), 1/2 tbsp olive oil, 1/2 cup chicken broth (125ml), 1/2 tsp sweet paprika

Steps:
1. Skin garlic and chop finely. Rinse lamb chops and pat dry. Sprinkle with salt and pepper and marinate for 5 minutes.
2. Wash and cut onion into wedges. Set aside.
3. Heat pan with a touch of butter and 1/2 tsp of oil. Sauté onion until fragrant.
4. Sprinkle sweet paprika over onion and stir well. Add in red wine and cook for a while. Pour in chicken broth and cook until sauce is done.
5. Pan-fry lamb chops until both sides turn golden brown. Pour in (4) onto the lambs. Cook for a while and serve.

材料：
羊扒2件、番茄1個、洋葱半個、蒜頭2瓣、鹽1/2茶匙、黑胡椒1/4茶匙

汁料：
油1/2湯匙、糖1/4茶匙、紅酒半杯（125ml）、橄欖油1/2湯匙、鮮雞湯半杯（125ml）、甜紅椒粉1/2茶匙

做法：
1. 蒜頭去皮、切茸；羊扒洗淨抹乾，加入鹽及胡椒粉醃5分鐘；
2. 洋葱洗淨，切角備用；
3. 燒熱少許牛油及1/2茶匙油，下洋葱爆香；
4. 加入甜紅椒粉炒勻後，倒入紅酒略煮，再倒入清雞湯煮成汁；
5. 羊扒放油鍋煎至兩面香黃，拌入（4）的汁料，略煮片刻即成。

 營養師提提您 Dietitian's Tips

羊肉含有豐富的鐵質，有助身體製造紅血球，很適合產後媽媽食用。

Lamb meat is rich in iron, an essential nutrient for the production of red blood cell. It is particularly good for postpartum mothers.

營養餸菜篇

營養成份 Nutrition information

碳水化合物(Carbohydrate)	☑	脂肪(Fat) 奧米加3脂肪酸(Omega-3 fatty acids)： 三文魚(salmon)	☑
蛋白質(Protein)	☑	礦物質(Minerals) 磷質(Phosphorous)	☑
維他命(Vitamins)	☑	纖維(Dietary Fibre)	☐

三文魚粒炒松子仁
Stir-fired Salmon Cubes with Pine Nuts
(2人份量 for 2 persons)

坐月必食
茶飲
補粥
燉品
補湯
飯餸

Ingredients:
120g salmon, 2 tbsp pine nut, 2 tbsp chopped celery, 3 Chinese mushrooms (diced), 1 tsp grated garlic, 1 tsp grated ginger

Sauce:
3 tbsp water, $1/2$ tsp oyster sauce, small amount of sugar and sesame oil, $1/2$ tsp corn flour

Steps:
1. Mix the sauce and set aside.
2. Cut salmon into thick dices. Stir-fry in oil until cooked. Take out salmon and set aside.
3. Heat 1 tsp of oil. Sauté garlic until fragrant. Add in pine nuts, chopped celery and Chinese mushrooms, and stir-fry thoroughly. Add salmon cubes into and stir well.
4. Pour sauce over ingredients and let boil until sauce is thickened. Serve.

材料：
三文魚肉120克、松子仁2湯匙、西芹（切碎）2湯匙、冬菇（切粒）3朵、蒜茸、薑茸各1茶匙

汁料：
水3湯匙、蠔油$1/2$茶匙、糖、麻油各少許、生粉$1/2$茶匙

做法：
1. 將汁料拌勻，備用；
2. 三文魚肉切粗粒，放入油鍋炒熟，盛起備用；
3. 燒熱1茶匙油，炒香蒜茸，放入松子仁、西芹及冬菇炒勻，再加入三文魚粒炒勻；
4. 下汁料煮滾至稠身後即成。

營養師提提您 Dietitian's Tips
三文魚所含有的奧米加3脂肪酸有助抗抑鬱，哺乳媽媽身體吸收後令母乳也含有奧米加3脂肪酸，有助嬰兒視力及腦部發展。

Salmon is high in Omega-3 fatty acids that help to prevent postpartum depression. For breast-feeding mothers who eat salmon, her breast milk also contains Omega-3 fatty acids, which are good for baby's eyesight and brain development.

營 養餸菜篇

營養成份 Nutrition information

碳水化合物(Carbohydrate)	☐	脂肪(Fat)	☐
蛋白質(Protein)	☐	礦物質(Minerals)	☑
		鐵(Iron)：豬肝及菠菜(pig's liver & spinach)	
		鈣(Calcium)：菠菜(spinach)	
維他命(Vitamins)	☑	纖維(Dietary Fibre)	☐
維他命A(Vitamin A)：豬肝及菠菜(pig's liver & spinach)			
葉酸(Folic acid)：菠菜(spinach)			

豬肝炒菠菜
Stir-fried Spinach with Pig's Liver
(2人份量 for 2 persons)

Ingredients:
80g pig's liver, 300g spinach, 3 slices fresh ginger

Seasonings:
1. ½ tbsp soy sauce, 1 tbsp rice wine, 1 tbsp corn flour
2. ¼ tsp salt, 1 tsp sugar

Steps:
1. Peel ginger and cut into slices. Rinse Chinese onion and chop finely. Wash spinach and cut into sections. Set aside.
2. Scald pig's liver for 30 minutes. Take out and cut into slices. Add in seasoning (1) and marinate for 5 minutes.
3. Heat 1 tbsp of oil. Stir-fry pig's liver slices over high heat until it is browned. Take out and set aside.
4. Add in spinach and stir-fry briefly. Re-add livers into pan and pour in seasoning (2) Stir well and serve.

材料：
豬肝2兩、菠菜300克、薑3片

調味料：
1. 生抽½湯匙、米酒1湯匙、粟粉1湯匙
2. 鹽¼茶匙、糖1茶匙

做法：
1. 薑去皮切片；葱洗淨，切成幼粒；菠菜洗淨，切段備用；
2. 豬肝浸泡30分鐘，取出切片，加入調味料（1）醃5分鐘；
3. 燒熱1湯匙油，放入豬肝以大火炒至變色，盛起；
4. 加入菠菜略炒一下，豬肝回鍋，並加入調味料（2）炒勻即可。

營養師提提您 Dietitian's Tips

菠菜和豬肝均含有豐富鐵質和維他命A，有助維持身體的免疫力。

Spinach and pig's liver are rich in iron and vitamin A, favourable for strengthening of immunity.

坐月必食

茶飲

補粥

燉品

補湯

飯餸

159

營養成份 Nutrition information

碳水化合物(Carbohydrate)	☐	脂肪(Fat)	☐
蛋白質(Protein)	☑	礦物質(Minerals)	☐
維他命(Vitamins)	☑	纖維(Dietary Fibre)	☑
維他命A、C (Vit. A, C)			

三色椒田雞
Stir-fried Frogs with Tri-colour Peppers
(2人份量 for 2 persons)

Ingredients:
120g edible frog, 1 green pepper, 1 red pepper, 1 yellow pepper, 2 cloves garlic, 1/2 tbsp fermented black bean

Seasonings:
1 1/2 tsp soy sauce, 2 tsp Shao Xing wine, 1/2 tsp corn flour, 1/4 cup water, 1 tsp sesame oil

Steps:
1. Remove the skin and all internal organs of edible frogs. Cut into pieces. Season with wine and pepper. Coat well and set aside.
2. Seed tri-colour peppers and cut into chunks. Rinse and set aside.
3. Peel garlic and rinse fermented beans. Mix garlic and beans together and crush in a bowl.
4. Heat pan with oil. Briefly stir-fry green, yellow and red peppers. Add in oil and water and stir-fry until tender. Set aside.
5. Heat pan with oil again. Sauté garlic and fermented black bean. Add in edible frog pieces. Briefly stir-fry, add seasoning. Cover and simmer for a while. When cooked, re-add the tri-colour peppers. Stir well and serve.

材料：
田雞120克、青椒1隻、紅椒1隻、黃椒1隻、蒜子2瓣、豆豉1/2湯匙

調味料：
生抽1 1/2茶匙、紹酒2茶匙、粟粉1/2茶匙、水1/4杯、麻油1茶匙

做法：
1. 田雞去皮和內臟，切件，加少許酒和胡椒粉拌勻，備用；
2. 青椒、紅椒、黃椒去籽、切塊，洗淨備用；
3. 蒜子去衣、豆豉洗淨，放入碗中用刀柄搗爛；
4. 燒熱油鍋，放入青椒、紅椒、黃椒略炒，加入少許油及水炒至剛熟，取出備用；
5. 再燒熱油鍋，爆香蒜茸豆豉，加入田雞，爆炒數下，下調味料，蓋上鍋蓋略焗，熟透後加入已炒熟的三色椒炒勻即成。

營養師提提您 Dietitian's Tips
三色甜椒含有豐富抗氧化物，例如胡蘿蔔素和維他命C，有助增強免疫系統。
Tri-colour peppers contain antioxidant nutrients such as carotene and vitamin C, which help strengthening of immunity.

坐月必食

茶飲

補粥

燉品

補湯

飯餸

營養餸菜篇

營養成份 Nutrition information

碳水化合物(Carbohydrate)	☐	脂肪(Fat)	☑
		單元不飽和脂肪(Mono-unsaturated fat)： 夏威夷果仁(macadamia nuts)	
蛋白質(Protein)	☑	礦物質(Minerals)	☐
維他命(Vitamins)	☑	纖維(Dietary Fibre)	☑
維他命A(Vit. A) 葉酸(Folic acid)：蘆筍(asparagus)		蘆筍(asparagus)	

蘆笋果仁炒帶子
Stir-fried Scallops with Asparagus
（2人份量 for 2 persons）

Ingredients:
80g scallop, 400g fresh asparagus, 40g macadamia nuts, suitable amount of carrot slices, 2 tsp shredded ginger

Seasonings:
$1/2$ tbsp soy sauce

Steps:
1. Rinse scallops. Wash asparagus and cut diagonally into sections. Set aside.
2. Blanch asparagus, scallops and carrot slices. Drain dry and set aside.
3. Heat pan with oil. Sauté shredded ginger until fragrant. Add in blanched vegetables and scallops and briefly stir.
4. Add in soy sauce and stir-fry until the sauce become thick. Add in macadamia nuts and serve.

材料：
鮮帶子80克、鮮蘆笋400克、夏威夷果仁40克、紅蘿蔔片適量、薑絲2茶匙

調味料：
生抽$1/2$湯匙

做法：
1. 帶子洗淨；鮮蘆笋洗淨，切斜段，備用；
2. 鮮蘆笋、帶子和紅蘿蔔片汆水，瀝乾水份，備用；
3. 燒熱油鍋，爆香薑絲，放入已汆水的蔬菜和帶子略炒；
4. 加入生抽，炒至汁乾，加入夏威夷果仁即成。

營養師提提您 Dietitian's Tips
帶子脂肪含量極低，配合含豐富單元不飽和脂肪酸的夏威夷果仁，大大減少飽和脂肪的攝取量。
Scallop contains low-fat and macadamia nuts are rich in mono-unsaturated fat. So the consumption of saturated fat can be greatly reduced.

坐月必食

茶飲

補粥

燉品

補湯

飯餸

營養餸菜篇

營養成份 Nutrition information

碳水化合物(Carbohydrate)	☐	脂肪(Fat)	☑
		低脂(Low Fat)：蝦(shrimps)	
蛋白質(Protein)	☐	礦物質(Minerals)	☐
維他命(Vitamins)	☐	纖維(Dietary Fibre)	☑

百合山藥炒蝦球
Stir-fried Shrimps with Lily Bulbs and Yams
(2人份量 for 2 persons)

Ingredients:
40g fresh lily bulb, 40g fresh Chinese yam, 50g celery, 100g shrimp, 2 tsp chopped ginger

Seasonings:
a pinch of salt

Steps:
1. Rinse lily bulbs and separate into petals. Wash Chinese yams and cut into thin slices.
2. Rinse celery and cut into sections of 3cm length. Shred ginger.
3. Rinse and blanch shrimps in boiling water briefly.
4. Heat pan with oil. Add in shrimps, lily bulbs, Chinese yams and celery pieces and stir well. Season with a pinch of salt. Stir-fry until cooked and serve.

材料：
鮮百合40克、山藥（鮮淮山）40克、西芹50克、蝦100克、薑茸2茶匙

調味料：
鹽少許

做法：
1. 百合洗淨，分成瓣狀；山藥洗淨，切薄片；
2. 西芹洗淨，切段（約3厘米）；薑切絲；
3. 蝦仁洗淨後，用沸水略灼；
4. 燒熱油鍋，加入蝦仁、百合、山藥，西芹炒勻，下少許鹽調味，炒熟即成。

營養師提提您 Dietitian's Tips
蝦的脂肪含量極低，且含豐富硒質，有助媽媽增強免疫系統。
Shrimp contains low-fat and is rich in selenium, which help strengthening of immunity.

營養成份 Nutrition information

碳水化合物(Carbohydrate)	☐	脂肪(Fat)	☐
蛋白質(Protein)	☑	礦物質(Minerals)	☐
維他命(Vitamins)	☐	纖維(Dietary Fibre)	☐

燕窩炒蛋白
Stir-fried Edible Bird's Nest with Egg White
(2人份量 for 2 persons)

Ingredients:
1 soaked edible bird's nest, 4 egg whites, small amount of chicken broth

Seasonings:
a pinch of salt

Steps:
1. Whisk egg white clockwise until foamy.
2. Slightly blanch edible bird's nest. Drain and set aside.
3. Heat pan with oil. Stir-fry edible bird's nest until done. Add in chicken broth and egg white. Stir quickly and season with salt. Stir-fry until dry and it is done.

材料：
浸發燕窩1盞、蛋白4隻、清雞湯少許

調味料：
鹽少許

做法：
1. 蛋白順時針方向打至發起；
2. 燕窩浸好，汆水，瀝乾備用；
3. 燒熱油鍋，下燕窩炒至熟，加清雞湯和蛋白快炒，再加入鹽調味，炒至乾身即可。

營養師提提您 Dietitian's Tips
蛋白含有豐富蛋白質，並脂肪含量為零。
Egg white are rich in protein and without any fats.

坐月必食

茶飲

補粥

燉品

補湯

飯餸

營養餸菜篇

營養成份 Nutrition information

碳水化合物(Carbohydrate)	☐	脂肪(Fat)	☐
蛋白質(Protein)	☑	礦物質(Minerals)	☐
維他命(Vitamins)	☐	纖維(Dietary Fibre)	☐

四色炒鱔絲

Stir-fried Finless Eel Shreds

(2人份量 for 2 persons)

Ingredients:
450g finless eel, 4 Chinese mushrooms (steamed and shredded), 2 tsp grated garlic, 1 red chili (shredded), 120g chives (cut into sections)

Marinade:
$1/4$ tsp salt, $1/2$ tsp starch flour, small amount of sesame oil and pepper

Sauce:
2 tbsp chicken broth, $1/2$ tsp sugar, $1/2$ tsp starch flour, $1/3$ tsp dark soy sauce

Steps:
1. Rub finless eel with salt and starch flour to remove sticky stuff. Rinse and cut into thick shreds. Marinate for 10 minutes.
2. Heat oil in a pan. Briefly stir-fry eel pieces until golden brown. Take them out and set aside.
3. Heat 1 tbsp of oil. Sauté shredded Chinese mushroom, red chili shreds, eel pieces and grated garlic until fragrant. Drizzle with wine.
4. Pour in sauce and stir-fry until well mixed. Pour chives and mix, and it is done.

材料：
黃鱔450克、冬菇（蒸熟、切絲）4朵、蒜茸2茶匙、紅椒1隻（切絲）、韭黃（切段）120克

醃料：
鹽$1/4$茶匙、生粉$1/2$茶匙、麻油、胡椒粉少許

汁料：
上湯2湯匙、糖$1/2$茶匙、生粉$1/2$茶匙、老抽$1/3$茶匙

做法：
1. 用鹽及生粉洗擦黃鱔去潺，洗淨，切粗條，加醃料拌勻，醃10分鐘；
2. 燒熱油，放入鱔條略炒至表面金黃色，盛起備用；
3. 下油1湯匙，下冬菇絲、紅椒絲、鱔條及蒜茸爆炒，濽酒；
4. 加入汁料炒勻，再加入韭黃拌勻即成。

營養師提提您 Dietitian's Tips
菜式顏色豐富，可增強媽媽的食慾。
This dish is colourful, which can enhance appetite.

坐月必食 茶飲 補粥 燉品 補湯 飯餸

營養餸菜篇

營養成份 Nutrition information

碳水化合物(Carbohydrate)	☐	脂肪(Fat)	☐
蛋白質(Protein)	☐	礦物質(Minerals)	☐
維他命(Vitamins)	☑	纖維(Dietary Fibre)	☑

脆脆雜錦藕片
Crispy Lotus Root Slices with Assorted Vegetables
(4人份量 for 4 persons)

Ingredients:
240g lotus root, 3 fresh Chinese mushrooms (halved), 6 slices carrot, 120g sugar snap pea, 4 pieces baby corn (halved), 1 tsp finely chopped garlic, 1 tsp finely chopped ginger

Sauce:
$1/2$ cup soup-stock, $1/4$ tsp salt, $1/2$ tsp sugar, 1 tsp oyster sauce

Steps:
1. Peel lotus root and cut into thin slices. Soak in water with a touch of white vinegar for 30 minutes. Take them out and blanch. Set aside.

2. Heat up oil in a non-stick pan and sauté sugar snap peas. Add in garlic, ginger, baby corns and carrot slices. Stir-fry until fragrant.

3. Add in fresh Chinese mushrooms and mix well with other ingredients. Add in lotus root slices.

4. Drizzle with wine and pour sauce over vegetables. Cover and simmer for a brief moment. Add in starch water to slightly thicken the sauce.

材料：
蓮藕240克、鮮冬菇（開邊）3朵、紅蘿蔔片6片、蜜糖豆120克、珍珠筍（開邊）4條、蒜茸1茶匙、薑茸1茶匙

汁料：
上湯$1/2$杯、鹽$1/4$茶匙、糖$1/2$茶匙、蠔油1茶匙

做法：
1. 蓮藕去皮，切薄片，放入已加少許白醋的清水中浸泡30分鐘後，取出汆水，備用；

2. 燒熱易潔鑊後下油，加入蜜糖豆爆炒後，加入蒜粒、薑粒、珍珠筍及紅蘿蔔片炒香；

3. 下鮮冬菇炒勻，放回蓮藕片；

4. 潲酒，加入汁料焗煮片刻，最後加生粉水埋薄芡即成。

營養師提提您 Dietitian's Tips
蔬菜含有豐富纖維素，有助防止媽媽們便秘。
Vegetables are rich in dietary fibre, helping to prevent constipation.

營養成份 Nutrition information

碳水化合物(Carbohydrate)	☑	脂肪(Fat)	☐
蛋白質(Protein)	☑	礦物質(Minerals)	☑
		鉀(Potassium)：栗子(chestnut)	
維他命(Vitamins)	☐	纖維(Dietary Fibre)	☐

日式栗子雞塊
Chicken with Chestnuts in Japanese Style
(4人份量 for 4 persons)

Ingredients:
120g chicken fillet, 150g chestnut, 20g ginger, 20g Chinese onion

Seasonings:
1 tbsp cooking wine, $\frac{1}{2}$ tbsp dark soy sauce, suitable amount of rock sugar and salt

Steps:
1. Rinse chicken and cut into pieces. Mix well with a touch of soy sauce.

2. Peel away shell and skin of chestnuts, set aside; cut and press the ginger; cut Chinese onion into sections.

3. Heat pan with oil. Stir-fry chicken pieces until golden brown. Take them out and set aside.

4. Heat pan over high heat. Sauté ginger and Chinese onion until fragrant. Add in chicken pieces and seasoning. Pour in suitable amount of water just to cover the chicken. When boiled, turn to low heat and stew until chicken is medium cooked to done. Add in chestnuts and continue to stew.

5. When chestnuts are softened, return to high heat and simmer the stew until sauce is reduced. Add in starch water to thicken the sauce and serve.

材料：
雞扒1件（約120克）、栗子150克、薑20克、葱20克

調味料：
料酒1湯匙、老抽$\frac{1}{2}$湯匙、冰糖、鹽適量

做法：
1. 雞扒洗淨、切方塊，加少許生抽拌勻；

2. 栗子去殼及衣，備用；薑切塊，拍鬆；葱切段；

3. 燒熱油鍋，下雞塊炒至金黃色，取出備用；

4. 大火燒熱油鍋，爆香薑葱，放入雞塊，加入調味料及適量清水（蓋過雞塊）煮沸後，轉小火燜至7成熟，加入栗子繼續燜；

5. 栗子煮至半熟時，改用大火將汁收乾，加入生粉水勾芡即成。

營養師提提您 Dietitian's Tips
栗子含豐富碳水化合物及鉀質，而且低脂肪。
Chestnut is rich in carbohydrates and potassium, and low in fat.

坐月必食

茶飲

補粥

燉品

補湯

飯餸

全彩食譜系列

(金牌)營養師的糖尿病甜美食譜

作者：張翠芬(註冊營養師)、
　　　林思為(註冊營養師)
頁數：192頁全彩
書價：HK$88、NT$350

(金牌)營養師的抗膽固醇私房菜

作者：張翠芬(註冊營養師)、
　　　林思為(註冊營養師)、
　　　劉碧珊(註冊營養師)
頁數：172頁全彩
書價：HK$88、NT$350

營養進補坐月食譜

作者：徐思濠(註冊中醫師)、
　　　胡美怡(註冊營養師)
頁數：176頁全彩
書價：HK$88、NT$390

營養師素食私房菜

作者：林思為(註冊營養師)、
　　　簡婉雯(註冊營養師)
頁數：160頁全彩
書價：HK$78、NT$390

營養師低卡私房菜

作者：林思為(註冊營養師)、
　　　黃思敏(註冊營養師)
頁數：168頁全彩
書價：HK$78、NT$390

排毒美容中醫湯水

作者：徐思濠(註冊中醫師)
頁數：136頁全彩
書價：HK$78、NT$299

(金牌)營養師的瘦身私房菜

作者：張翠芬(註冊營養師)、
　　　林思為(註冊營養師)
頁數：160頁全彩
書價：HK$98、NT$390

(金牌)營養師的小學生午餐便當

作者：張翠芬(註冊營養師)、
　　　林思為(註冊營養師)
頁數：144頁全彩
書價：HK$78、NT$350

0-2 歲快樂寶寶食譜 & 全方位照護手冊

作者：張翠芬(註冊營養師)、
　　　林思為(註冊營養師)、
　　　鄭碧純(兒科專科醫生)
頁數：160頁全彩
書價：HK$68、NT$299

懷孕坐月營養師食譜

作者：張翠芬(註冊營養師)、
　　　林思為(註冊營養師)、
　　　簡婉雯(註冊營養師)
頁數：160頁全彩
書價：HK$68、NT$350

輕鬆抗癌營養師食譜

作者：基督教聯合那打素社康服務(註冊營養師)
頁數：208頁全彩
書價：HK$78、NT$299

營養師的輕怡瘦身甜品

作者：基督教聯合那打素社康服務(註冊營養師)
頁數：136頁全彩
書價：HK$78、NT$390

《營養進補坐月食譜》（修訂第四版）

編　　著：徐思濠(註冊中醫師)
營養顧問：胡美怡(註冊營養師)
封面設計：SiyuanCreation
版面設計：跨版生活製作部、麥碧心
攝　　影：傅穎鈿
翻　　譯：廖惠堂、葉子菁、關小春
責任編輯：伍喜碧、高家華

出版：跨版生活圖書出版
地址：荃灣沙咀道11-19號達貿中心211室
電話：31535574　　傳真：31627223
專頁：http://www.facebook.com/crossborderbook
網頁：http://www.crossborderbook.net
電郵：crossborderbook@yahoo.com.hk

發行：泛華發行代理有限公司
地址：香港新界將軍澳工業邨駿昌街7號星島新聞集團大廈
電話：2798 2220　　傳真：2796 5471
網頁：http://www.gccd.com.hk
電郵：gccd@singtaonewscorp.com

台灣總經銷：永盈出版行銷有限公司
地址：231新北市新店區中正路499號4樓
電話：(02)2218 0701　　傳真：(02)2218 0704

印刷：鴻基印刷有限公司

出版日期：2020年10月第5次印刷
定價：HK$88　NT$390
ISBN：978-988-75022-3-4

出版社法律顧問：勞潔儀律師行

©版權所有 翻印必究

本書所刊載之網頁畫面和商標的版權均屬各該公司所有，書中只用作輔助説明之用。本書內容屬於作者個人觀點和意見，不代表本出版社立場。跨版生活圖書出版（Cross Border Publishing Company）擁有本書及本書所刊載的文字、插圖、相片、版面、封面及所有其他部分的版權，未經授權，不得以印刷、電子或任何其他形式轉載、複製或翻印，本公司保留一切法律追究的權利。

免責聲明
跨版生活圖書出版社和本書作者已盡力確保圖文資料準確，但資料只供一般參考用途，對於有關資料在任何特定情況下使用時的準確性或可靠性，本社並沒有作出任何明示或隱含的陳述、申述、保證或擔保。本社一概不會就因本書資料之任何不確、遺漏或過時而引致之任何損失或損害承擔任何責任或法律責任。

(我們力求每本圖書的資訊都是正確無誤，且每次更新都會努力檢查資料是否最新，但如讀者發現任何紕漏或錯誤之處，歡迎以電郵或電話告訴我們！)